Conflict and Change in Mexican Economic Strategy

Conflict and Change in Mexican Economic Strategy

Implications for Mexico and for Latin America

John Sheahan

Printed with the assistance of the Tinker Foundation

Monograph Series, 34
Center for U.S.-Mexican Studies
University of California, San Diego

1991

Printed in the United States of America by
the Center for U.S.-Mexican Studies
University of California, San Diego

ISBN 1-878367-05-6

Contents

List of Tables and Figures	vii
Abstract	ix
Introduction	1
Macroeconomic Issues: Conflicts among Goals	7
Inflation, Real Exchange Rates, Exports, and Wages	15
Liberalization and the Structures of Trade and Production	35
Liberalization, Wage Policies, Poverty, and Income Distribution	45
Orientation and Character of the Mexican State	55
Some Implications and Questions for Latin America	67
Conclusions	73
References	77
About the Author	82

Tables and Figures

TABLES

1. Real Resource Transfers into and out of Mexico and Five Other Latin American Countries, 1980–1989   9

2. Changes in GDP Per Capita and in Wages, 1980–1989   10

3. Rates of Inflation for Selected Years from 1982 to 1990   13

4. Mexican Exports of Manufactured Products, 1980–1989, as Reported by the Bank of Mexico and by the World Bank   20

5. Real Exchange Rates and Exports of Manufactures for Mexico and Five Other Latin American Countries, Selected Years from 1985 to 1989   25

FIGURES

1. Indexes of GDP Per Capita, Average Real Wages in Manufacturing, and Minimum Real Wages, 1980–1989   11

2. Real Exchange Rates and the Rate of Inflation, 1980–1990   19

3. Changes in the Real Exchange Rate and in Average Real Wages in Manufacturing, 1981–1989   29

Abstract

Mexico's process of adjustment to the debt crisis evolved from a combination of macroeconomic restraint plus export promotion into a thoroughgoing redirection of the country's prior economic strategy. This study uses comparisons to other Latin American countries to bring out distinctive characteristics of Mexico's path, and to Chile and Colombia in particular as examples of different kinds of economic liberalization, with different economic and social consequences.

Mexico's adjustment has been distinctive in terms of success in promoting industrial exports and in escape from a process of increasing inflation, but also in terms of the degree to which real wages were driven down. The key policy conflict has been between the use of devaluation to turn the industrial sector outward toward competition in export markets, and concern for the negative effects of devaluation on both inflation and real wages.

The main questions considered with respect to liberalization are its effects on the structures of production and trade, on poverty and income distribution, on the capacity of the state to guide the character of development, on the traditionally inclusionary character of the society, and on the possibilities for more complete democracy. Liberalization can have many different meanings. Mexico's version has moved toward priority for efficiency and for reliance on private investors and market forces, relative to concern for social goals or for the kinds of corporatist negotiation so effective in the past in holding the society together. This version has been determined by executive decision; its future character may depend to a high degree on whether or not popular preferences can be effectively expressed through a more democratic political system.

Introduction

Latin American countries have changed their economic strategies in fundamental ways in response to the debt crisis of the 1980s, and Mexico perhaps more thoroughly than any other country in the region. What began as a severely puritanical adjustment to correct macroeconomic disequilibrium, a straightforward effort to hold down domestic spending and achieve external balance, turned into a striking effort to restructure the Mexican economic system as a whole, to get away from the weaknesses thought to underlie both the debt problem and the long-term difficulties of modernizing the economy.

The essence of the restructuring effort has been to liberalize the economy, to reject the system of protection and state-led industrialization on which Mexico has relied since the 1930s (Cárdenas 1987, 1990; Villarreal 1988). The central question in this study is what the consequences of this effort have been so far and where they may be leading. On one level, this is a question of adjustment to the debt crisis, of degrees of success or failure in correcting macroeconomic disequilibrium. On a deeper level, it is a question of the kind of economy, and society, that is being shaped by the new economic strategy.

Much of this discussion compares the process of economic adjustment in Mexico in the 1980s to those of Chile and Colombia, as examples of prior redirection toward more open economic systems, and to Argentina, Brazil, and Peru as examples of markedly different strategies during the 1980s. Measures of adjustment in this decade bring out two quite different paths, one for Chile and

---

I would like to express my thanks to Robin King and to Kevin Middlebrook for very useful comments on drafts of this study, and to Carlos Bazdresch, Brígida García de Garza, Gustavo Garza, Georgiana Kessel, Flavia Rodríguez de Grey, Ignacio Trigueros, Víctor Urquidi, and Norma Rocío and René Villarreal for helpful discussions of these issues.

Colombia and a second for Argentina, Brazil, and Peru. A striking characteristic of Mexico's adjustment path is that the results resemble the latter group at first, with similarly low success, but then they begin a trajectory of change from the initial pattern toward that of Chile and Colombia.

Comparisons with Chile and Colombia may also help illuminate broader implications of Mexico's changes toward a more outwardly oriented economic system: Colombia for its experience with a new emphasis on promoting industrial exports and a moderate degree of liberalization starting in 1967, and Chile for its much more sweeping liberalization after the military coup in 1973. The two reorientations of economic strategy were implemented in different ways with different goals: in Chile under intense repression and with little concern for heightened inequality, or perhaps even a goal of raising property income relative to labor; in Colombia under a more nearly democratic system, with a considerable effort to improve conditions for the poor. Clearly there are different ways to go about transformation toward a more open economic system, and room for different consequences.

Mexico's process of adjustment in the 1980s started from a much greater degree of macroeconomic imbalance than those of the other countries, except Chile. The government's strategy ran into sharp conflict on two closely related fronts: (1) efforts to restrain demand and inflation by severe fiscal contraction clashed with efforts to redirect production toward exports by aggressive devaluation; and (2) traditional concerns for holding the society together by protection of wages and by extensive social programs came up against the contrary effort to maintain tight fiscal austerity and increase the industrial sector's competitiveness in external markets. The economic side of these conflicts shows up clearly in the erratic course of the real exchange rate, alternately driven up to stimulate an export-led recovery and then allowed to fall in order to lessen inflation and downward pressure on real wages. The political-social conflicts were in effect embedded in this macroeconomic indecision: increases in the real exchange rate succeeded in encouraging rapid growth of industrial exports but at the same time drove down real wages, while fiscal restraint succeeded in freeing resources for export growth but drastically cut social programs.

Trade liberalization began on a limited scale in 1985, in the midst of these unresolved conflicts, possibly in response to external pressures but also as a support for that side of the government's strategy directed toward making Mexican industry more competitive. It reinforced the effect of real devaluation in encouraging exports, but, unlike devaluation, trade liberalization also helped to restrain inflation. At the same time, it increased pressures to hold down labor costs and to withdraw support from both minimum wages and remaining social programs, all of which helped to build up a strong new basis of political opposition threatening the entrenched control of the political system by the Partido Revolucionario Institucional (PRI).

Interactions of economic liberalization and political liberalization raise a host of treacherous questions. The redirection of Mexican economic strategy can be viewed as a liberating movement that could lead to a more productive economy and more sustainable growth, that might lessen the high degree of inequality characteristic of Mexico in the past, and that might lead to a more open political system through escape from state control of the economy. Or it can be viewed as a sellout of everything distinctly Mexican about the country's economic system, as a retreat from any goal of independent industrialization, and as an abandonment of traditional efforts to use state intervention to shape the character of development in a more inclusionary style, with more support for labor and peasants, than would have been likely in a system left up to market forces. Both views can be supported in some respects, the first perhaps more in terms of gains for efficiency and more sustainable growth, the second more in terms of negative social consequences. These tensions suggest conflict rather than mutual support between the goal of a fully liberalized economic system and the goal of a more fully open and fully accepted democracy: the economic goal favors pressures to protect a liberal economy by continued restrictions on full democracy, while realization of the political goal would almost surely lead toward severe qualification, or conceivably rejection, of economic liberalization.

Two of the most penetrating analyses of relationships between economic strategies and political outcomes in Latin America—by Carlos Díaz Alejandro (1983) and Albert Hirschman (1979)—provide useful warnings against any sense of *necessary* links between economic liberalization and democracy. Both suggest that the connections between economic strategies and political outcomes are too diffuse, and too variable in response to the particular historical context, to allow any confident presumption that such a change in economic strategy is either favorable or unfavorable for democracy. Granting the strength of their arguments, it still seems more than an accident that the most aggressive changes toward economic liberalization in the 1970s, in the three Southern Cone countries, were all imposed by repressive military governments shortly after overturning democracies, as if liberalization in some sense required suppressing popular choice. The same kind of state violence and suppression of democratic institutions was true of Brazil's turn toward a more market-determined and less protected economic system after its military coup in 1964, but it was not true of Colombia's milder turn toward a more export-oriented strategy from 1967. Four cases in which varying kinds of economic liberalization were adopted as the economic strategies of repressive military governments do not prove a necessary association, but they certainly suggest problems about the connections between economic and political liberalization. One exception in which a mild version of economic liberalization proved to be possible within the context of democratic institutions (or at least of the somewhat qualified democracy of Colombia at the time), though it does not disprove the reality of such problems, at least supports the possibility of a happier outcome.

The main hypothesis in this discussion is that the exceptional tensions between economic and political liberalization in Latin America can be explained in part by the fact that the existing economic structures of most of these countries make the consequences of reliance on market forces far more adverse for equality than is the case in the northern industrialized countries. Under structural conditions in which market forces work to spread the gains of economic growth widely, including the great majority of the population, they are likely to be not only accepted but favored by a democratic political process. Under structural conditions in which they work adversely for the majority, or even for a substantial minority, then a democratic political process would be expected to reject or to qualify seriously any economic strategy relying on market forces.

The difficulties of achieving more complete democracy in modern Latin America, and of keeping it in place in periods of strain, have many possible explanations, but surely one important strand of the problem is that the kinds of liberal economic policies favorable for efficiency and more stable growth are so often unfavorable for lower-income people. The existing structures of most of these economies could make the consequences of liberalization so adverse to so many people that genuinely open elections by an informed electorate would consistently require severe restrictions on the kind and degree of liberalization that could be maintained (see Sheahan 1991).

Mexico's economic liberalization has not been forced through by repression of the Southern Cone variety, but neither has it been a product of any fully open democratic system. It has instead been associated with a touch-and-go series of changes toward democracy, marked by important increases in the acceptance and roles of opposition political parties but also by the absence of effective constraint on the use of state resources to support the traditional ruling PRI party and by recurrent examples of falsified election results used to keep power in the hands of the governing party. Economic liberalization in the context of this still unresolved struggle for genuine political liberalization raises intriguing questions in two directions: how the former may affect the latter, and how the character of economic liberalization might evolve in the event that it is held in place by a political system itself not fully open to expression of public preferences.

While the central concern of this study is the process of adjustment and liberalization in Mexico, the framework in terms of comparisons to the experiences of other Latin American countries is intended to bring out implications for them as well. Relatively successful liberalization in Mexico could have profound effects on perceptions and choices in the region as a whole. Likely influences are a more general and more definitive turn away from heterodox or populist macroeconomic strategies, reduced possibilities for common strategies for such purposes as debt renegotiation or control of foreign investment, and changing possibilities for democracies in the region.

Changes from liberal economic systems to periods of active state intervention, and back again, seem to go in long cycles, as the costs of one extreme build up pressures to go too far in the opposite direction. Much of the region went from overly generalized economic liberalization in the half century up to the 1930s, in political contexts of narrowly circumscribed or nonexistent democracy in most countries, to excessive and often ill-conceived kinds of intervention in the first postwar decades, and now back toward what could become a socially costly kind of liberalization. But the new cycle is not obliged by history or economics to go overboard once again in favor of reliance on markets and incentives for private investors. If it includes more meaningful democracy than in Latin America's past, public preferences could exert pressure for more effective concern to reshape these economies in ways favorable for wider participation in the gains of economic growth.

Macroeconomic Issues: Conflicts among Goals

When the debt crisis hit in 1982, Mexico's immediate problems were similar to those of the majority of Latin American countries and the initial responses were much alike. The first requirement was to get out of the preceding disequilibrium, characterized by an excess of spending over production and a corresponding excess of imports over exports. The sudden interruption of external financing forced Mexico, along with the other deeply indebted countries of Latin America, to put the brakes on domestic demand and created strong pressure for currency devaluations in order to shift demand away from imports and to encourage exports. Within these common reactions each country had considerable room for differences in emphasis, reflecting its own internal balance of forces and preferences. Mexico's pattern of response stands out for the intensity of initial corrective measures, for a clear-cut move toward more conservative solutions just when others opted for more unorthodox experiments, and finally for an extraordinary reversal of prior economic strategy.

Degrees of success or failure in adjustment need to be considered in many different dimensions. Whatever the nation's particular priorities, to concentrate on any one set of goals usually implies sacrificing performance in other directions. Compared to other leading Latin American countries, the Mexican governments after 1982 gave greater priority to restoring macroeconomic balance and promoting new exports, at the cost of greater decreases in wages, output per capita, and standards of living. Concern for inflation seems to have gone back and forth in conflict with concern for promoting exports. Inflation was allowed to rise steeply as part of the initial adjustment process in 1983, was curbed in 1984-85, shot up again in 1986-87, and was then brought back under better control after 1987. The key conflict on this level of economic policy has been over exchange rates: priority for checking inflation argues in favor of keeping the nominal exchange rate stable, but priority for promotion of exports and concern for ability to compete with imports in a more open economy both argue in favor of devaluation in order to keep the real exchange rate either steady or rising.

The initial combination of tight restraint on demand and strong devaluation offered both a hope and a threat. The hopeful side of the combination is that restraint on domestic demand may reduce inflationary pressures while devaluation helps keep up production by raising sales to external markets. If devaluation works well it can lead to a higher-level external balance based on export growth rather than a lower-level balance relying on contraction. The threat is that this combination of measures can go in the wrong direction on all counts: that restraint of domestic demand may lead to falling output uncompensated by any significant increase of exports, while devaluation drives up domestic prices and inflationary expectations despite falling demand and output. For Mexico, as for the majority of Latin American countries, the threat initially proved more realistic than the hope. Output fell but inflation speeded up, setting off a dangerous race between further devaluation intended to offset inflation and further inflationary consequences of devaluation itself.

In a long-run perspective these objectives are mutually consistent: restoration of macroeconomic balance could reduce the threat of inflationary consequences from export promotion and open the way to a more sustainable growth path, with both rising employment and rising real wages. But trying to do everything at once, starting from deep initial disequilibrium, can be explosive. If a country gives priority to export promotion through devaluation, this can touch off runaway inflation unless domestic demand and wages are driven down mercilessly, and sometimes even if they are. If priority is given instead to stopping inflation, as in Bolivia in 1985, it becomes practically imperative to suspend rapid devaluation and export promotion. Jeffrey Sachs, the key architect of the Bolivian stabilization program, argues that it is a costly mistake to follow the path that Mexico chose; that it is essential for adjustment programs to concentrate on establishing macroeconomic balance, for a period long enough to eliminate inflationary expectations, before launching any major effort at export promotion (Sachs 1987). Mexico seemed for a time to fit that argument in the sense that it tried to do both at once and suffered highly inflationary consequences. But in the perspective of the last decade as a whole it may have gained some important advantages from that effort. Mexico seems to be coming out of the inflationary trap in a better position with respect to exports and structural change, though not as yet with respect to wages and output, than the countries that tried either more heterodox measures or the policy balance suggested by the Bolivian model of stabilization.

The drastic nature of Mexico's initial efforts to cut back demand and stimulate exports can be explained in part by the degree of its initial disequilibrium. Prior disregard of excess spending gave the country an exceptionally high external deficit as of 1980, worsened by weakening exports in 1981. As shown in table 1, real resource transfers into Mexico relative to its GDP (the excess of spending over output) were much higher than for the region as a whole. For 1981, Mexico's net use of external resources was equal to 11.3 percent of GDP, nearly double the corresponding ratio for the region. Of the five other countries included for comparison in table 1, four (all but Chile) show degrees of disequilibrium much below that of Mexico.

TABLE 1
REAL RESOURCE TRANSFERS INTO AND OUT OF MEXICO AND FIVE OTHER LATIN AMERICAN COUNTRIES, 1980–1989
RESOURCE TRANSFERS AS PERCENT OF GDP
(POSITIVE BALANCE IS INFLOW, NEGATIVE IS OUTFLOW)

	1980	1981	1982	1983	Change 1981–83	Average 1983–88	Estimate 1989
Mexico	9.9	11.3	0.6	−6.2	−17.5	−4.8	0.1
Argentina	6.9	5.6	−0.7	−1.8	−7.4	−2.3	−6.4
Brazil	2.3	−0.2	0	−2.4	−2.2	−3.6	−4.9
Chile	4.7	10.8	−1.9	−5.7	−16.5	−6.4	−4.0
Colombia	3.1	5.4	6.9	5.2	−0.2	0.8	−3.0
Peru	3.9	6.9	6.6	2.2	−4.7	0.1	−3.2
Latin America	5.9	5.8	2.6	−1.8	−7.4	−2.3	−3.2

Source: Inter-American Development Bank, *Economic and Social Progress in Latin America, 1990 Report* (Washington: Johns Hopkins University Press for the IDB, 1990), 21.

The severity of the initial Mexican cutback was correspondingly extreme. The inflow of resources in 1981 was replaced by an outflow equal to 6.2 percent of GDP in 1983, a change that cut back domestic resource use relative to output by 17.5 percent of GDP. That was far more drastic than the cutback for the region as a whole or for any of the other countries compared except Chile.

The administration of Miguel de la Madrid maintained tight austerity, holding domestic resource use below production by an average of 4.8 percent of GDP for the six years from 1983 through 1988. Again, the only closely similar experience, with even tighter restraint, is that of Chile. In contrast, the first full year of the administration of Carlos Salinas (1989) marked a major change. For the first time since 1981, Mexican economic policy stopped holding domestic demand below output. It was the only one of the countries in table 1 that did so: it switched from the conservative to the expansionary end of the distribution. Provisional estimates for 1990 suggest almost exactly the same balance, with a net resource inflow of less than 1 percent of GDP.[1]

Table 2 gives intercountry comparisons of changes in output per capita and wages for 1980–1989, and figure 1 charts year-by-year changes for Mexico. As late as 1989 Mexico's GDP per capita was down 9 percent compared to its 1980 level, average real wages in manufacturing were down 24 percent, and real minimum wages were down 47 percent. The cut in real minimum wages was far steeper than in any other country except Peru and far greater than the fall in real income per capita. Of course, the per capita supply of goods and services fell much more than income: comparing 1989

TABLE 2
CHANGES IN GDP PER CAPITA AND IN WAGES, 1980–1989
(PERCENTAGE CHANGES FROM BASE TO END YEAR)

	GDP per capita 1980–89	Real minimum wages in urban employment 1980–89	1987–89	Real average wages in manufacturing 1980–89	1987–89
Mexico	−9	−47	−17	−24[a]	4[a]
Argentina	−26	−32	−43	−28	−18
Brazil	0	−31	6	7[b]	5[b]
Chile	11	−37	10	3	9
Colombia	12	5	−7	19	0
Peru	−25	−77	−61	−58	−59
Latin America	−9	−25	−12	n.a.	n.a.

[a]Average wages in manufacturing in Mexico for 1989 are estimated from data for January–October.

[b]Average wage in manufacturing in Brazil is measure for Rio de Janeiro. Wage increase for São Paulo were much higher for both periods but then wages for both of them show sharp decreases in 1990: the index for Rio fell 20 percent between 1989 and 1990.

Sources: GDP and minimum wages from Inter-American Development Bank, *Economic and Social Progress in Latin America, 1990 Report* (Washington: Johns Hopkins University Press for the IDB, 1990), 28 and 265; average wages in manufacturing from Economic Commission for Latin America and the Caribbean, *Preliminary Overview of the Economy of Latin America and the Caribbean, 1990* (Santiago: ECLAC, December 1990), 27.

[1] The provisional estimate for 1990 is based on ECLAC 1990: table 15. The net capital inflow for 1990 is estimated at $8.9 billion, compared to an outflow of $7.9 billion for interest and profit remittances.

FIGURE 1

Indexes of GDP Per Capita, Average Real Wages in Manufacturing, and Minimum Real Wages in Urban Employment, Mexico, 1980-1989

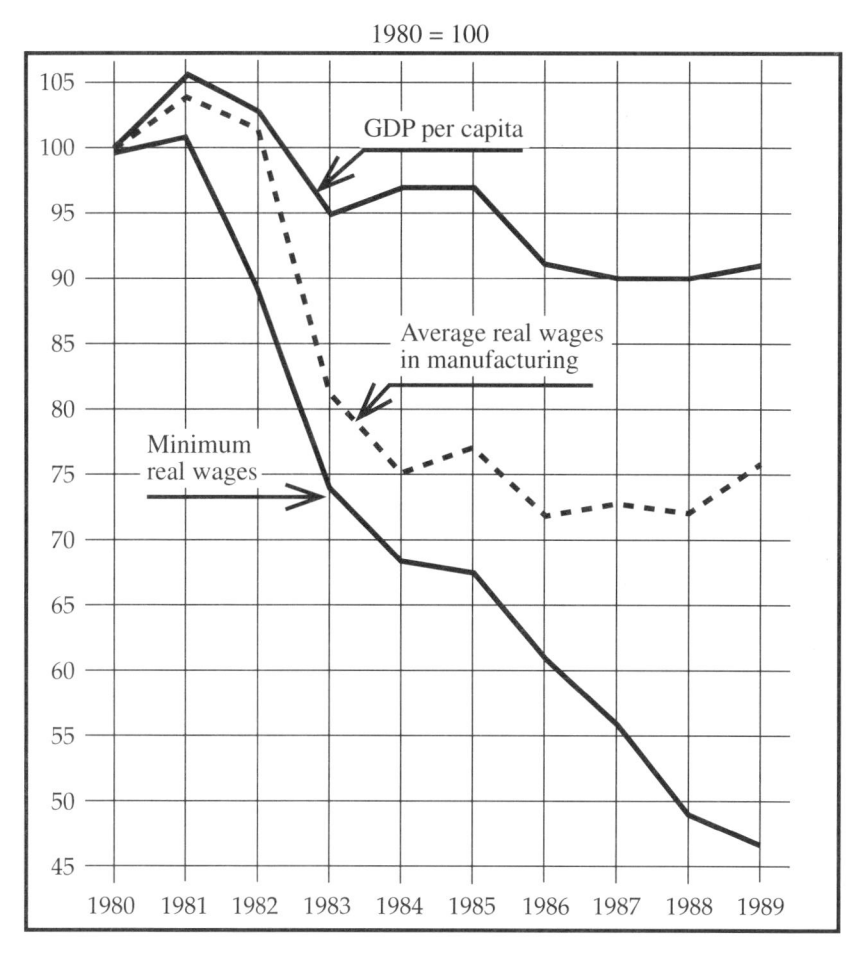

1980 = 100

Sources: GDP per capita and minimum real wages from Inter-American Development Bank, *Economic and Social Progress in Latin America*, 1990 Report (Washington, D.C.: Johns Hopkins University Press for the IDB, 1990), 28 and 265; average real wages in manufacturing from Economic Commission for Latin America and the Caribbean, *Preliminary Overview of the Economy of Latin America and the Caribbean, 1990* (Santiago: ECLAC, December 1990), 27.

to 1980, it went down by 10 percent of GDP because of the elimination of the prior resource inflow from abroad, on top of the drop of 9 percent in output per capita. The combined drop in the supply of goods and services per capita was still a great deal less than the cut in real minimum wages and also less than the cut of 24 percent in average wages in manufacturing.

It is possible that earnings of the rural labor force and in the urban informal sector did not go down as sharply as formal-sector wages. Estimates of the proportion of rural families below the poverty line show a decrease in 1986 as compared to 1970, from 49 to 43 percent, while the percentage for urban families rose from 20 to 23 percent (Feres and León 1990: table 6). Merilee Grindle suggests that earnings in the rural sector were helped by a rise in the share of agriculture in GDP from 1983 to 1986 and that rural living standards were also supported by continuing or possibly even increasing remittances from emigrants (Grindle 1991). It was the wage earners who bore the brunt: while the real minimum wage came down steeply, the percentage of families earning less than two minimum wages increased from 47 percent in 1981 to almost 60 percent in 1987 (Solís 1990: 45). National accounts data indicate a sharp change in the distribution of income between property ownership and labor. Total wage payments relative to income from ownership of capital fell steeply, from 0.64 in 1980 to 0.40 in 1988.[2]

The fall in Mexico's output per capita between 1980 and 1989 just matched that for the region as a whole. The decrease was much less than in Argentina or Peru, but it contrasts with rising output per capita in both Chile and Colombia. Real minimum wages were cut twice as steeply in Mexico as they were for the region as a whole. The fall in average wages in manufacturing was comparable to that in Argentina and much less than in Peru, though it again contrasts with actual increases in Chile and Colombia, as well as Brazil. By these measures, Mexico's adjustment process had unusually high costs in terms of reduced wages. It should be noted that the countries that did best in terms of both output and wages were Chile and Colombia—those with the most nearly open economies at the start.

In terms of ability to bring down inflation, the Mexican record looked at first dangerously weak but subsequently turned very much for the better (see table 3). Its rate of inflation was far higher than in Chile or Colombia in 1982, even prior to its leap in 1983, though it was still well below Argentina and Brazil in both of these

---

[2]King 1990: 17. Returns to capital are those estimated in Mexican national accounts; they do not include returns on the very substantial Mexican-owned assets held abroad.

TABLE 3
RATES OF INFLATION FOR SELECTED YEARS FROM 1982 TO 1990

	1982	1983	1987	1988	1989	1990
Mexico	59	102	132	114	20	27
Argentina	165	344	131	343	3079	2314
Brazil	98	142	230	682	1287	2968
Chile	10	27	20	15	17	26
Colombia	25	20	23	28	26	29
Peru	64	111	86	666	3399	7482

Source: Inter-American Development Bank, *Economic and Social Progress in Latin America, 1991 Report* (Washington: Johns Hopkins University Press for the IDB, 1991), individual country statistical pages.

years. Mexico's inflation slowed down in 1984–85 but then rose to a record high in 1987. By that year the countries included in table 3 had split into two distinct groups: Chile and Colombia continued to hold inflation below 25 percent, while Peruvian inflation had risen to 86 percent and the other three, including Mexico, were well over 100 percent. Then came the crucial change: Mexican inflation came down to the levels of Chile and Colombia, while inflation in the other three countries continued to rise. On this criterion of adjustment, Mexico switched from the losing track to the relatively successful group.

Inflation, Real Exchange Rates, Exports, and Wages

The key measures that brought Mexico's inflation rate down temporarily in 1984–85, and more decisively from 1987, were decisions in both periods to stop devaluing the currency at rates equal to or greater than the rate of inflation, the move toward trade liberalization, and the negotiation in 1987 of a form of incomes policy under the Economic Solidarity Pact.

Inflation leaped upward in 1982–83 and again in 1986–87, the two periods in which the real exchange rate was greatly raised, and then came down swiftly in 1984–85 and after 1987 when it was allowed to appreciate. Appreciation clearly served as a brake on inflation but it also threatened one of the most striking positive achievements of the 1980s, the impressive growth of the country's exports of manufactures. From 1987, the goal of curbing inflation was given priority over that of export promotion, despite the latter's central role in determining the prospects for sustained growth.

The real exchange rate can be measured in many different ways, but for all of them a basic question is whether or not, in the presence of inflation, the currency is devalued rapidly enough to prevent the inflation from destroying incentives to export.[3] The measure of the real rate used here, as published by the Inter-American Development Bank, is the ratio of foreign to domestic prices when both are expressed in the same currency: an index of the nominal exchange rate multiplied by the ratio between an index of foreign prices and the index of domestic consumer prices. With a given nominal exchange rate, domestic inflation reduces the ratio between foreign and domestic prices and thereby drives down the

---

[3]See Edwards 1989 for a thorough discussion of alternative measures of real exchange rates and their economic significance. Cottani, Cavallo, and Khan 1990: 61–76, includes empirical tests of relationships between real exchange rates and characteristics of macroeconomic performance, as well as useful further discussion of their interpretation.

real exchange rate. Such a fall in the real exchange rate both undermines the competitive position of exporters and raises incentives to import rather than buy from domestic suppliers. Exactly that process, allowing inflation to drive down the real exchange rate, has been perhaps the most frequent direct cause of foreign exchange crises and aborted economic growth in postwar Latin America.

The presidency of José López Portillo started with a competitive real exchange rate, established by devaluation to deal with the balance-of-payments crisis of 1976. But the rise in domestic spending and inflation associated with the oil boom and the government's rising fiscal deficits again drove the real exchange rate down, aggravating Mexico's external debt and the costs of the following debt crisis. The government of Miguel de la Madrid recognized the disastrous effect of overvaluation, determined to correct it, and did. The government of Carlos Salinas surely recognizes the problems bound to result from overvaluation, most especially for an economy in process of liberalization, but has been giving troublesome signs of falling back into it.

The main reason that so many governments have been reluctant to devalue in order to protect the real exchange rate when it is pulled down by inflation, even when the negative effects on exports and the trade balance become clear, is that it almost invariably aggravates the inflation. It does so directly by raising the prices of tradable goods and indirectly by raising costs of production in all activities using tradable inputs and by reinforcing inflationary expectations. The opposite course, allowing the real exchange rate to become overvalued, provides temporary help in restraining inflation at the cost of piling up greater future trouble.

Deliberate overvaluation may also have specific policy objectives in its own right. It has often been associated both with import substitution, as a way to keep down the costs of imported capital equipment for industrialists, and with the efforts of populist governments to raise real wages. Appreciation of the currency in real terms can in fact help raise real wages temporarily, until its negative effect on exports fosters external collapse (Sachs 1989). But populists have no monopoly on the practice of overvaluation. Perhaps the greatest surprise of the 1970s in this respect was to see the ultraconservative government of General Pinochet in Chile, and the similarly conservative military junta in Argentina from 1976, turn to the familiar populist technique with exactly the same negative effects on external balance.

In both of these cases, the basic purpose of holding down the nominal price of foreign exchange was to slow inflation. In Chile, the promising recovery from 1976 to 1979 was set back badly by a decision to freeze the nominal exchange rate in 1979, allowing the real rate to appreciate. The policy came very close to success in stopping inflation, but the other side of the coin was that exports fell, imports rose swiftly, and the Chilean debt rapidly grew to unmanageable proportions. It is interesting to see in table 1 above that Chile's combination of liberalization with tight fiscal policy— with no budget deficit at all—failed to prevent national spending from overshooting production to almost exactly the same degree as in Mexico in 1981. In Mexico, the main explanation of the unsustainably high current account deficit was overspending driven by the budget deficit and the oil boom; in Chile both of those factors were missing but currency overvaluation accomplished the same destructive result (Corbo, de Melo, and Tybout 1986; Edwards and Edwards 1987; Ramos 1986).

Mexican economic policy prior to the 1980s gave high value to the avoidance of devaluation: a stable currency (in nominal terms) was presented to the public repeatedly as a demonstration of successful management of the economy. The nominal exchange rate was held fixed from 1954 to 1976. This stability proved to be compatible with low external deficits through the decade of the 1960s because of the country's remarkable success in holding down inflation. The average annual rate of inflation in this decade was only 2.7 percent (IMF 1983). But as inflation went up to an average of 16.6 percent a year for the 1970s, a fixed nominal exchange rate began to mean a rapidly appreciating real rate. That appreciation in real terms, combined with fast growth of domestic spending and demand for imports, created rising external deficits that proved to be unsustainable by 1976. The government was forced to turn to the International Monetary Fund for help, agreeing to devaluation as one of the conditions for new credit from the Fund.

After the devaluation of 1976 the government returned to the basic policy of a fixed nominal exchange rate for as long as it could. With the oil boom and the accompanying illusion that the country was free of foreign exchange constraints, both government and private spending increased rapidly. So did the demand for imports and the price level. Even with the concurrent growth of earnings from oil exports, spending on imports greatly outpaced export revenue. Mexico had to borrow abroad to finance the accelerating deficit on current account. That was at first very easy to do. Foreign banks rushed to provide loans—in some cases, they actively pushed loans on Mexico—in the overshooting behavior typical of private capital markets during periods of overoptimism.

From 1980, the government began making small changes in the nominal exchange rate, though not enough to keep up with the rate of inflation. The real exchange rate continued to deteriorate until the debt crisis hit in 1982. Faced with an overwhelming current account deficit and a total reversal of behavior by the international banks, from previous overlending to a paralysis of new credit, the government then resorted to drastic devaluation along with contractionary measures to drive down domestic demand. The real exchange rate rose rapidly in 1982 and 1983, pushing up the rate of inflation despite the domestic contraction and setting the country off on the difficult course of trying to restore a viable real exchange rate while also trying to check inflation.

Real exchange rates and inflation proved to be synchronized to a high degree in the 1980s (see figure 2). Except for 1981, every increase or decrease in the real exchange rate is matched by a corresponding increase or decrease in the rate of inflation. A simple regression of the annual rate of inflation on the percentage change of the real exchange rate gives a statistically significant relationship in which a 1 percent increase in the exchange rate is associated with an increase of 1.14 percentage points in the rate of inflation (an increase of 1.7 percent in the rate of inflation, evaluated at its mean).[4]

This close association is by no means a logically necessary connection; it is an unusual characteristic of Mexican experience beginning in 1982. The more common historical pattern is that of 1981: the two measures go in opposite directions when a rise in inflation from causes other than devaluation is not offset by a compensating change in the exchange rate. This more frequent inverse relationship is especially clear for Peru in the 1980s. Inflation there accelerated in the second half of the decade, driven mainly by rapid increases in central bank credit creation, but the government resisted compensating devaluation for fear that this would make the inflation even worse: the real exchange rate fell to half its level in 1985.[5]

The parallel movements of inflation and the real exchange rate in Mexico from 1982 through 1989 indicate that other possible causes of inflation were kept under control. The most important factor brought under control was the primary fiscal deficit, excluding interest payments from consideration. The primary deficit had reached 9.5 percent of GDP in 1981 and remained at 9.4 percent in 1982, but in 1983 it was wiped out and replaced by a surplus equal

---

[4] For the period 1980 through 1989, the adjusted r square of the regression is 0.63. The standard error of the parameter estimate for the real exchange rate is 0.28, with a t statistic of 4.03.

[5] See table 3 on the acceleration of inflation, and table 5 on the corresponding fall in Peru's real exchange rate.

FIGURE 2

Real Exchange Rates and the Rate of Inflation, 1980-1990

Index of real exchange rate (1980 = 100)

Inflation (average percent increase of consumer prices)

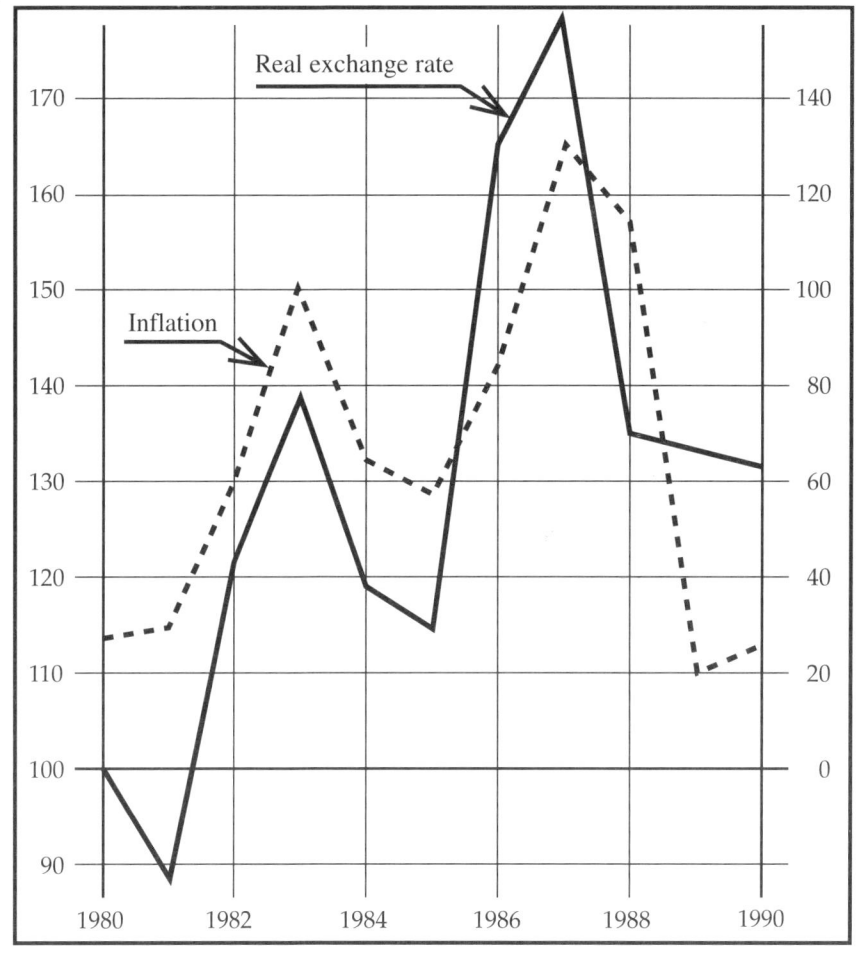

Real exchange rate

Inflation

Source: Inter-American Development Bank, *Economic and Social Progress in Latin America, 1991 Report* (Washington D.C.: Johns Hopkins University Press for the IDB, 1991), 147.

to 3.3 percent of GDP. The primary fiscal balance was then stabilized at approximately this ratio to GDP.[6] The result of this effective restraint on the fiscal side was that, for this specific period, changes in the real exchange rate became the main determinant of changes in the rate of inflation.

The government responded to the rising rate of inflation in 1982–83 by slowing up devaluation, allowing the real exchange rate to fall by 17 percent between 1983 and 1985. That brought inflation down from 102 to 58 percent. The cost was a strong brake on the growth of industrial exports. As shown in table 4, exports of manufactures as reported by the Bank of Mexico fell 11 percent in 1985, reversing the prior trend of rapid growth. The parallel series for Mexican exports as reported by the World Bank, adjusted for international comparability, shows continuing growth but at a much lower rate than in the two preceding years. If Mexican economic policy had been dominated by greater concern for inflation than for economic transformation, or if a new oil export boom or significant debt relief had taken off the pressure for adjustment, the success of 1985 in cutting inflation might have led to a decision to stop devaluation indefinitely. Instead, severely adverse external

TABLE 4
MEXICAN EXPORTS OF MANUFACTURED PRODUCTS, 1980–1989, AS REPORTED BY THE BANK OF MEXICO AND BY THE WORLD BANK
(MILLION DOLLARS)

	As reported by the Bank of Mexico	As reported by the World Bank
1980	3,030	2,234
1981	3,361	2,664
1982	3,018	1,804
1983	4,583	3,287
1984	5,595	4,249
1985	4,978	4,640
1986	7,116	7,126
1987	9,907	9,774
1988	11,523	11,577
1989	12,530	11,592

Sources: Bank of Mexico, *Economic Indicators*, November 1989, table IV-H-5 for 1980–86, and March 1991, table IV-2 for 1987–89; World Bank, *World Tables 1991* (Baltimore: Johns Hopkins University Press for the World Bank, 1991), 407.

[6] Alberro and Cambiaso 1989: table 8. The financial deficit, including interest payments, was correspondingly cut but still remained a high share of GDP: it was brought down from 17.6 percent of GDP in 1982 to 9.0 percent in 1983 and held at about this level until it rose again in 1986.

conditions acted to heighten concern for trade policy.[7] The de la Madrid administration opted for a return to real devaluation, this time with a new emphasis on liberalizing the economy to force Mexican industry to become more competitive.

Liberalization would mean the practical elimination of much of the industrial sector if the real exchange rate were persistently overvalued: it could become impossible for even the most efficient industries to compete. The prior system of thorough protection meant that the industrial sector could grow along with the domestic market, but liberalization changes the situation radically: only those firms able to match or undersell foreign competitors can survive. The great hope for the success of the liberalization strategy, at least in terms of continuing industrial growth, is that enough firms will prove able to compete in external markets, and grow through exports, to offset the inescapable elimination of those firms unable to get their costs down enough to be competitive. But the possibilities both for export success and for competition against imports depend on keeping the real exchange rate high enough to give Mexican firms a decent chance.

The relationships between exchange rate policies and the growth of industrial exports have been a long-standing matter of debate, in Mexico as in the rest of Latin America. Traditional preferences for discretionary economic management, as opposed to reliance on incentives acting through the price system, put the emphasis on the use of selective intervention as the main means to promote new exports. Mexico followed this course with some success in the 1970s and could conceivably revert to it if liberalization is rejected in the future. In this view, the surge of industrial exports achieved in the 1980s is not explained by increases in the real exchange rate but instead by preceding industrial policies. This position has been argued especially with respect to the record of the automobile industry's transformation from pure import substitution to the most dynamic industrial exporter of the 1980s.[8]

The opposing interpretation is that incentives provided by the real exchange rate can stimulate industrial exports without the need for selective intervention and without the possible distortions that such intervention may foster. A possible bridge between the two views is that both selective intervention and an appropriate real exchange rate can work together to give a far better performance of industrial exports, and of the economy as a whole, than

---

[7]See King 1990 on the systematic connection between changes in degrees of external pressure and the pattern of domestic policy choices.

[8]Mercado and Taniura 1990; Twomey 1990. Mercado and Taniura do not even mention the exchange rate as a possible factor affecting automobile exports; Twomey explicitly argues for the dominant role of selective intervention (pp. 1–2).

can be expected from relying exclusively on either one. If incentives through the exchange rate were allowed to fall so low as to eliminate profitability of exports, selective intervention would be most unlikely to give powerful results. But the capacity of industries to respond to increases in the real exchange rate by increasing exports can almost surely be improved by selective intervention to that end: it seems clear that the boom of Mexican industrial exports in the 1980s owes a good deal to such prior intervention.

The case for emphasis on selective intervention gains support from the performance of Mexican industrial exports through the decade of the 1970s. Despite the use of a fixed nominal exchange rate from 1970 to 1976, exports of manufactures rose three times over in this period, from $321 million to $1,012 million (World Bank 1991b: 406). The peak was reached in 1974 and export growth did not resume until 1978, but the total then shot up to $1,708 million and continued growing despite a deteriorating real exchange rate until 1982.

Selective export promotion for industrial products started in the 1960s and took on a much more important role in the 1970s. It included two different kinds of methods: (1) subsidies for exports and (2) forms of industrial policy of the East Asian style, using administrative regulations and favors to push particular firms and industries into exporting.

The value of selective export subsidies for manufacturing exports has been estimated at 6 percent of the nominal exchange rate as of 1965, 9 percent by 1970, and fully 79 percent by 1975 (Behar 1988: table C.1). After the nominal devaluation of 1976, the value of these subsidies was at first cut sharply but it then started rising again to a second peak of 49 percent of the nominal exchange rate in 1981. The substantial real devaluations of 1982 and 1983 then allowed the subsidies to be cut down greatly, to less than 1 percent of the nominal exchange rate by 1983.

If the value of each year's level of subsidy is added to the nominal exchange rate (to permit estimation of a real exchange rate corrected to include the subsidies), the net effect for export incentives is highly interesting. This adjusted estimate of the real exchange rate shows a high degree of stability in the first half of the 1970s, rather than the apparent deterioration without such correction, and then shows a gradual rise. The seemingly mysterious rise of manufactured exports in the presence of a fixed nominal exchange rate becomes less puzzling: incentives for such exports were not in fact allowed to deteriorate. Jaime Behar's econometric tests of the supply response of manufacturing exports to changes in the real exchange rate, as corrected to include these subsidies, strongly support the hypothesis that these exports are systematically linked to the real exchange rate (Behar 1988: chaps. 4 and 5).

The second kind of intervention, through administrative regulations, may have played an important complementary role. The most important case has been the automobile industry—originally for exports of parts, then automobile engines, and then in the course of the 1980s for assembled passenger cars. The industry developed at first under protection for import substitution, accompanied by increasing pressures from the government to force the final assembly firms to buy their inputs from domestic suppliers. Then in 1969, in reaction to the companies' opposition to this pressure to use costly domestic inputs, the government switched to an alternative, using a foreign exchange budget to give the companies room to continue importing inputs if they earned the foreign exchange to pay for them through exports. In effect, part of the rent made possible by domestic protection was recaptured from the companies by requiring them to export even if the exports were not separately profitable.

Decrees in 1972 and 1977 carried the pressures further. The requirements for exports were strengthened but sensibly allowed companies to choose the particular products they could use for exporting. That meant that firms that wished to stay in the domestic market were given strong incentives to carry out investments on a sufficient scale to bring down their unit costs and cut losses on required exports, or even turn them into a new source of profits. The appeal of cooperation gained added weight from the announcements of Mexico's new oil discoveries and the resumption of a domestic boom: the expected value to the firms of being able to participate in a rising domestic market made it more worthwhile than ever to respond to the government's efforts to promote exports. The producers launched a major round of new investment projects from 1977, this time aiming at economies of scale and control of quality sufficient to support a combination of exports and rising domestic sales (Twomey 1990; Unger 1990: 137–51).

When the debt crisis hit in 1982, the dreams of rapidly increasing sales to the domestic market were abruptly deflated. But the new plants, with their up-to-date technologies and large capacity, provided a powerful base for expansion into export markets. That base might have led to significant new exports in any case, but it was soon given added stimulus from a change in administrative regulations combined with real devaluation in response to the crisis. The change in administrative rules, in 1983, allowed exporters to reduce required ratios of domestic to total inputs, in effect using a step toward liberalization as an inducement to increase exports (Mercado and Taniura 1990: 7–8). Between that measure and the accompanying real devaluation, the industry's exports went up by 23 percent in 1982, 99 percent in 1983, and 43 percent in 1984. They rose from 14 percent of total manufacturing exports in 1980 to 21 percent of the much larger total in 1984, and then to 29 percent by 1989 (Unger 1990: tables II.1 and A.II.1; Mercado and Taniura 1990: table 2).

Proponents of specific intervention to stimulate exports, as opposed to use of devaluation, can cite the example of the automobile industry as a demonstration of its positive effect in setting the stage for extraordinarily successful export growth in this industry. Proponents of generalized stimulus through the exchange rate can equally cite the fact that exports took off dramatically just when the real exchange rate was raised; they can add the further point that when the real exchange rate was allowed to fall again in 1984 and 1985, the industry's export growth slowed in 1984 and stopped completely in 1985 (Unger 1990: table A.II.1). A possible reply to that is that it was only the older component of the industry's exports that reacted negatively to the real appreciation in 1985; the newer exports of engines and assembled passenger cars kept right on growing (Mercado and Taniura 1990: fig. 1).

The oscillations of exchange rate policy and the trend of industrial exports in the 1980s are summarized for Mexico and for the other five reference countries in table 5. Mexico accomplished a strong devaluation in real terms in 1983, let the real exchange rate fall back again in the next two years, returned to real devaluation in 1986–87, and then let the real rate slide again after 1987. Chile and Colombia followed steadier paths toward a gradually rising real exchange rate from 1983 to 1988, though Chile then let its real rate fall in 1989. Brazil and Peru headed into deepening trouble at the end of the decade in many respects, aggravated by letting their real exchange rates fall.

Considering the trends for industrial exports, Mexico stands out for the most impressive record. As reported by the World Bank, industrial exports in 1989 were 5.2 times their 1980 level. Argentina, Chile, and Colombia all raised their real exchange rates and their industrial exports, while Peru let its real exchange rate fall and its exports of manufactures fell correspondingly. Brazil is out of line with the others in that its industrial exports did well up to 1988, despite using only a very modest real depreciation up to 1987 and then allowing a significant appreciation. The positive result for exports through 1988 probably owes a good deal to the extraordinary dynamism of Brazilian industry prior to the debt crisis, with a long period of policy conditions favorable for industrial exports from 1964 through the 1970s. If such favorable conditions are built into expectations by consistent policy support, then the real

TABLE 5
REAL EXCHANGE RATES AND EXPORTS OF MANUFACTURES FOR
MEXICO AND FIVE OTHER LATIN AMERICAN COUNTRIES,
SELECTED YEARS FROM 1985 TO 1989

	Real exchange rates[a] on base 1980 = 100				Industrial exports on base 1980 = 100		
	1985	1987	1988	1989	1987	1988	1989
Mexico[b]	115	179	150	130	438	518	519
Argentina	227	235	267	291	108	124	136
Brazil	99	105	97	74	151	208	183
Chile	145	185	198	192	100	236	114
Colombia	110	165	171	176	135	159	160
Peru	106	84	92	51	71	86	n.a.

[a]Real exchange rates are effective rates as reported by the Inter-American Development Bank, based on nominal exchange rates corrected for changes of consumer prices. A rise in the rate indicates a devaluation in real terms, increasing foreign prices relative to domestic.

[b]Data for Mexican exports do not include export shipments by maquiladoras. Their value added increased 3.5 times between 1980 and 1989, reaching $3 billion in 1989.

Sources: Real exchange rates from Inter-American Development Bank, *Economic and Social Progress in Latin America, 1990 Report* (Washington: Johns Hopkins University Press for the IDB, 1990), p. 31; exports of manufactures from country tables in World Bank, *World Tables 1991* (Baltimore: Johns Hopkins University Press for the World Bank, 1991). Data in note (b) for maquiladoras from Bank of Mexico, *Economic Indicators*, September 1990. It should be noted that the Bank of Mexico gives a considerably different picture for Mexico's exports of manufactures in 1989, showing an increase of 8.7 percent (see table 4).

exchange rate may not need to be raised greatly for sustained export performance; it only needs to be protected from falling for any length of time. The fall in Brazilian industrial exports in 1989 may have meant that the real appreciation after 1987 was beginning to be too much for even this country's dynamic industrial sector.

Year-by-year percentage changes of Mexico's exports of manufactures during the 1980s are not closely related to changes in the real exchange rate; changes in the latter are frequently followed by corresponding changes in exports the next year, but the correlation between the two series is not statistically significant. The dominant factor is the strong trend of export growth, surely aided by the higher average real exchange rate after 1981, but also due in part to promotional intervention and perhaps in even larger measure to the strong contraction of domestic markets. The only path open for profitable growth was to put a new emphasis on exporting. But the incentive to do so was magnified by more favorable real exchange rates in some years, and the growth trend slowed up abruptly in the first two intervals in which the rate was allowed to appreciate.

The preceding fast growth of manufacturing exports, which nearly doubled between 1976 and 1981, was stopped cold after the real exchange rate appreciated by 14 percent in 1981. As reported by the Bank of Mexico, exports fell 10 percent between 1981 and 1982. As reported by the World Bank, they fell by a third. That was the point at which the debt crisis forced radical changes in Mexican economic policies. The real exchange rate was raised 58 percent between 1981 and 1983. Exports of manufactures responded strongly, rising by more than 50 percent from 1982 to 1983. So far, so good: the supply response of Mexican industry proved to be powerfully positive. But the effect on inflation was equally quick and strong; it went from 28 percent in 1981 to 102 percent for 1983.

The jump in inflation led to a decision to slow devaluation in 1984 and 1985, allowing the real exchange rate to fall by 17 percent. That cut the rate of inflation nearly in half but again slowed the growth of industrial exports (or actually drove them down by 11 percent in 1985, according to the Bank of Mexico series). In the context of the decision to proceed with liberalization, the government sensibly chose to move the real exchange rate back up again, increasing it by 55 percent between 1985 and 1987. That paid off quickly by renewing the growth of manufacturing exports: they practically doubled between these two years. That renewed success on the side of export promotion certainly favored the intended restructuring of the economy, but it also once again set off higher inflation: from 58 percent for 1985 to a totally unacceptable 132 percent for 1987. Something clearly had to be done differently.

The two additional policies that helped to change the balance were an acceleration of the process of trade liberalization initiated in 1985 and the Economic Solidarity Pact negotiated at the end of 1987. Trade liberalization helped exporters through easier and cheaper access to imported inputs, and through increased competition acting to restrain domestic prices. For any given exchange rate, lower tariff protection should both lessen inflation and favor exports by reducing costs of production.

The other side of restraint on costs and prices, the Economic Solidarity Pact, can be seen as a form of incomes policy meant to resolve the conflict between stabilization and export promotion. The pact had three main economic components: (1) agreements by labor and industry to accept direct restraint of wages and prices; (2) agreement by the government to slow devaluation in order to hold down increases of costs and prices; and (3) insistence by the government, agreed to by industry, on much more extensive reductions in protection (Whitehead 1989). It was an affirmation of continued trade liberalization accompanied by extra-market intervention to restrain costs and prices.

The logic of the Solidarity Pact offered a possible way out of the conflict between attempts at stabilization and efforts to promote exports through devaluation. In effect, the pact was a decision to use administrative intervention—a key ingredient of the heterodox adjustment policies that failed in Argentina, Brazil, and Peru—in support of a policy of liberalization. The Mexican experience with such wage and price restraint proved distinctly more effective than in the other three cases, perhaps in part because of the traditions of Mexico's corporative system, but surely also because it was accompanied by tight fiscal restraint in a context of depressed aggregate demand. In the event, the pact and its included element of slowing down devaluation brought the rate of inflation down quickly, from 132 to 20 percent in one year. The new policy also allowed the real exchange rate to fall, by 27 percent from 1987 to 1989. According to the World Bank series for exports of manufactures, they continued growing well in 1988 but practically stopped rising in 1989. According to the Bank of Mexico, they continued growing in 1989 at 8.7 percent, and then nearly 10 percent further in the first eleven months of 1990 (Banco de México 1991: table IV-2).

Mexico's exceptional balancing act from 1987 to 1990, bringing down inflation without actually decreasing industrial exports, is perhaps the most positive evidence so far of possible success for renewed industrial growth in a more open economy. That success could be undermined by continued decreases in the real exchange rate, and in fact it was allowed to deteriorate further in 1990 and the first half of 1991. This could become a serious threat to Mexican industry in a liberalized economic system. But it remains a positive sign, as well as surprising, that these exports held up in 1989 (or even grew, according to the Bank of Mexico series), especially when contrasted to their downturns when the real exchange rate was allowed to fall in 1982 and 1984–85.

Continued strength of industrial exports despite the adverse trend of the real exchange rate could conceivably be explained by any one or a combination of several different factors in the new context of 1988–90: by yet further downward pressure on real wages under the Economic Solidarity Pact, by a high degree of preceding undervaluation as a result of the sharp depreciation of 1986–87, by benefits of liberalization on the side of production costs, or by a growing confidence on the part of industrialists in the likelihood that national economic policy will not be allowed to revert to an anti-export orientation.

Of these possible explanations, further wage repression seems the least likely candidate. The data available for average real wages in manufacturing from 1987 do not suggest that further wage repression under the Solidarity Pact was used to keep manufactured exports growing. On the contrary, they apparently turned back upward in 1989. They had been driven down 28 percent between 1980 and 1986 and were still at that low in 1988, but they then increased 4.8 percent in 1989 and 2.7 percent in 1990.[9] That picture is consistent with the hypothesis that the real exchange rate is itself an important determinant of real wages: when it was being raised to promote exports, and in the process stimulated inflation, the effect was to undercut the purchasing power of wages.

There was a fairly consistent negative relationship between changes in the real exchange rate and changes in average manufacturing wages from 1980 to 1989. To bring out the association graphically, in figure 3 increases in the real exchange rate are plotted against decreases in real wages. A simple regression shows a statistically significant negative relationship, with a 1 percent increase in the real exchange rate associated with a decrease of 0.49 percent in the index of real wages.[10] Small wonder that labor opposed continuing devaluation and sought, though unsuccessfully, a system of wage indexation. Failing such indexation, the Solidarity Pact offered perhaps the next best solution: an agreement by the government to slow down devaluation and allow the real exchange rate to rise, with the implication that real wages could be expected to stop falling.

This negative relationship between the real exchange rate and the course of average real wages seems clear enough for this period in Mexico, but it can be seriously misleading about longer-term possibilities. Two countries succeeded in raising real wages in the 1980–89 period: Chile by 3 percent and Colombia by 19 percent (see table 2). Both of these countries also raised their real exchange rates substantially in this period: Chile by 92 percent and Colombia by 76 percent (see table 5). Conversely, Peru let its real exchange rate appreciate greatly, and its real wages went down even faster than those in Mexico. What these three cases of positive association between real exchange rates and real wages suggest is that an adequate real rate favors more successful long-term growth, rising employment, and rising real wages, while an overvalued real rate handicaps growth and is more likely to be associated with stagnant or falling real wages because of poor macroeconomic performance (and in Peru's case, greatly falling real income per capita). That hypothesis raises a question for Mexican experience in the 1980s: why does the relationship go the other way?

---

[9] IDB 1991: 126.

[10] The adjusted r square is only 0.44 but the t statistic for the parameter estimate on the real exchange rate is -2.54.

FIGURE 3

Percentage Changes in the Real Exchange Rate and in Average Real Wages in Manufacturing, Mexico, 1980-1989

Percent decrease in real wages

Percent increase in real exchange rate

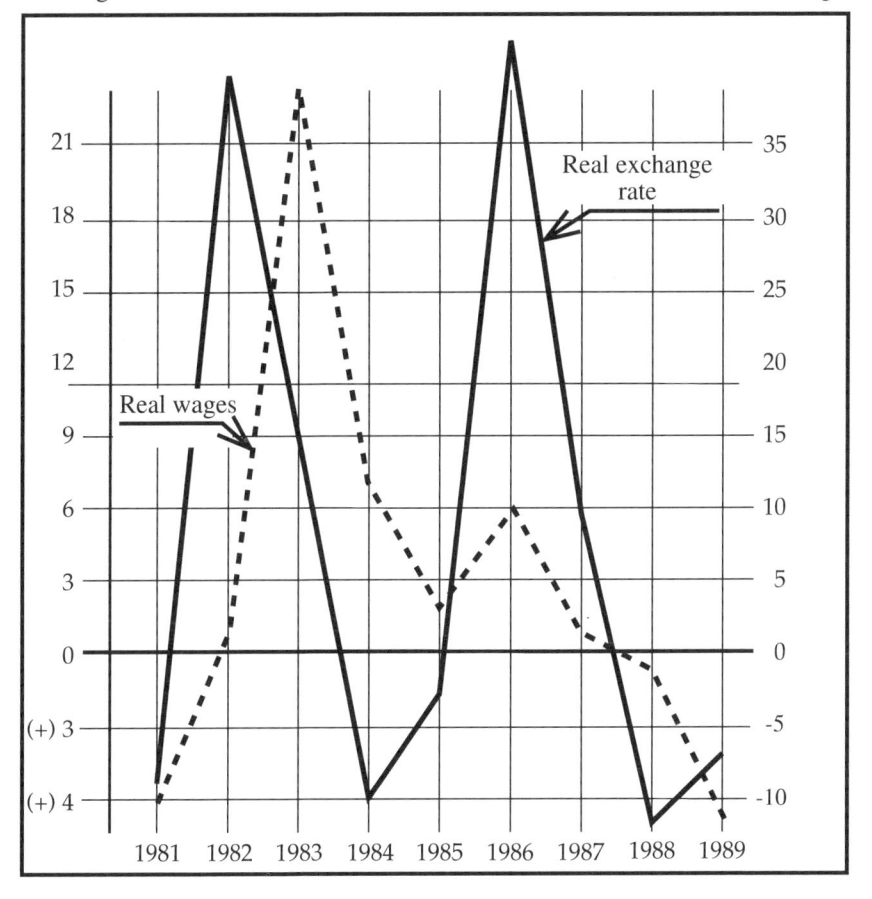

Real exchange rate

Real wages

Source: Real exchange rate from Inter-American Development Bank, *Economic and Social Progress in Latin America, 1991 Report* (Washington D.C.: Johns Hopkins University Press for the IDB, 1991), 126.

The answer to this question involves both national policy with respect to wage rates in manufacturing and consideration of the role of productivity change. On the level of national policy, decisions to raise the price of foreign exchange are often accompanied by agreements to raise money wages, in the realistic expectation that the devaluation will increase the price level. If money wages are raised by the same proportion as the price of foreign exchange, that pulls up domestic costs and prices by about the same percentage and wipes out any significant increase in the real exchange rate: the whole operation merely worsens inflation. Mexican policy in the 1980s went the other way, holding increases of money wages below the degree of devaluation, precisely in order to ensure a rise in the real exchange rate. The government's ability to do so depended both on a relatively passive reaction by organized labor, discussed further below, and on a depressed internal market with weak demand for labor. In Chile and in Colombia during these years, or more exactly from 1984 on, vigorous overall growth and rising employment favored rising money wages and the latter kept up with rates of devaluation: real wages and the real exchange rate rose together in favorable market conditions.

The second explanatory factor, the course of productivity change, is more fundamental. It is difficult to raise real wages for any prolonged period, regardless of the real exchange rate, faster than the rate of growth of output per worker. But when the period is one of major changes in the real exchange rate, these changes can react back on the relationship between productivity and real wages. The issue requires a distinction between the factors determining prices of tradable goods and those determining the prices of nontradable goods and services. Prices of tradable goods are in the usual case directly responsive to increases in the price of foreign exchange: they are likely to increase by approximately the same percentage as the currency is devalued. Prices of nontradable goods and services are not as directly affected. They are pulled up to some degree by the increases in costs of tradable inputs, but they are also greatly affected by wage costs, which are in turn determined by money wages and by productivity. If money wages go up but labor productivity does not, this pulls up the prices of nontradables at rates parallel to money wages, with the result that real wages go nowhere. But if productivity increases rapidly, money wages can rise faster than the prices of nontradables, permitting real wages to rise in terms of purchasing power over nontradables. Real devaluation pulls real wages down in terms of purchasing power over tradable goods, but this effect can be offset, permitting a rise in overall real wages, if they are rising fast enough in terms of nontradables. The Colombian example in the

1980s, in contrast to that of Mexico, illustrates this result particularly well.

In Colombia, gross output per worker in manufacturing increased 48 percent between 1980 and 1988 (World Bank 1991a: 216). If the real exchange rate had stayed constant and if the overall economy had performed as well as it did despite this fixed rate, real wages in manufacturing could have risen at approximately the rate of growth of productivity. But the overall performance of the economy depended on achieving a more competitive exchange rate: the real devaluation was a powerful aid to overall growth and therefore to favorable labor market conditions and to rising productivity in the first place. Considered separately, the real rate raised the prices of tradable goods more rapidly than productivity increased (the real exchange rate rose by 71 percent up to 1988, while the estimate for productivity increase was 48 percent). The basic rise in productivity permitted increasing real wages in terms of nontradable goods, but the real devaluation probably reduced them in terms of tradable goods. The net effect remained positive because the productivity growth, facilitated by real devaluation, was enough to offset the subtraction from real wages on the side of tradable goods.

The Mexican experience was markedly different, even though the same forces were at work in the same ways. Gross output per worker in manufacturing increased only 11 percent between 1980 and 1988, less than a fourth the gain in Colombia (World Bank 1991a: 217). Still, that increase in productivity would have permitted a gain in real wages in terms of nontradable goods. But real devaluation, by 45 percent between 1980 and 1988, took away too much purchasing power on the side of tradable goods, leaving the negative relationship between real exchange rates and real wages noted above. By the same token, real appreciation in 1988 and 1989 helped account for the apparent end of the fall in real wages and the slight rise in 1989. The way to restore rising real wages in the short run seems clear: keep on appreciating the real exchange rate. The problem with that response, if continued for any length of time, is that it would be very likely to choke off recovery and lead to falling real wages from macroeconomic failure, as it did in Peru.

The fact that Mexican industrial exports continued to grow in 1988–90 despite the real appreciation is not attributable to further wage repression. A more likely explanatory factor is that the 1986–87 devaluation was so drastic that it served to provide a safety margin in the following years. Exactly where the real exchange rate needs to go in order to ensure continued success with exports is an open question, but it seems probable that the real rate became higher than necessary in 1987. By the most traditional method of estimating overvaluation or undervaluation in terms of purchasing power parity—comparing changes in domestic and external prices from a base year considered to provide something like an equilibrium starting point—the exchange rate appears to have been overvalued by 21 percent in 1981 but undervalued by 45 percent in mid-1986.[11] By the IDB measure of the real exchange rate used in figure 2, the degree of undervaluation increased in 1987. After the fall in the real exchange rate in the next two years it was still, as of 1989, 48 percent above the level of 1981. That combination of estimates would suggest that the currency was still undervalued by something on the order of 25 percent in 1989. Further appreciation in 1990 cut into this remaining margin, raising more seriously the possibility of choking off the growth of industrial exports.

Purchasing power parity calculations can give useful clues but they do not provide a safe set of guidelines. If the key target is to keep industrial exports growing in order to make possible a more sustainable and more competitive process of development, then the test is not so much the position of relative prices as it is the performance of the exports. If the growth of industrial exports had stopped in 1989–90, as it did twice before in the course of the 1980s, that would be a powerful signal of a need to raise the real exchange rate. The Bank of Mexico's data on actual results for 1989–90 are reassuring in this respect: according to this series, both direct exports and value added in maquiladora sales continued to increase. The World Bank measure of exports in 1989, showing almost no growth for that year, is a more troublesome sign.

The government's policy orientation since 1987 suggests an understandable wish to keep working on inflation by holding back on devaluation, allowing the real exchange rate to continue falling. This preference may be gaining support from the conviction that foreign exchange is becoming less of an issue because of a revival of foreign investment. Why not let the real exchange rate appreciate if market forces work in this direction as a result of improving balance-of-payments conditions?

A particularly suggestive explanation of the market-based interpretation of exchange rate policy, by Javier Beristain and Ignacio Trigueros (1990), emphasizes that the government should not try to make the real exchange rate go anywhere in particular as a matter of deliberate choice; it should instead respect the pressures created by changes in the availability of foreign exchange. These authors criticize the de la Madrid administration for overdevaluing in 1987 when the adverse external conditions of 1986 were already turning for the better: the rise in inflation is seen as a

---

[11] Alberro y Cambiaso (1989: 177) use 1978 as their base year.

consequence of arbitrary intervention running against market forces. The argument suggests reliance on such forces, with careful attention to the overall balance of payments but with no separate concern for what happens to industrial exports.

This widely shared position is consistent with liberalization in placing trust in market forces as opposed to definition of specific goals for industrial policy. It could be consistent with industrial growth if foreign exchange earnings from primary products and services, and the inflow from borrowing and direct foreign investment, were so constrained that overall financial balance required a real exchange rate sufficiently high to make the industrial sector competitive in external markets. Under contrary conditions, with higher earnings from primary exports and services or a high capital inflow, it could result in atrophy of the industrial sector.

The kind of box that might have trapped the Mexican economy if the government had opted for a market-determined exchange rate, instead of following its discretionary targets, is made clear by comparison with Peru's experience when a new government there adopted a drastic adjustment program in August 1990. The country's real exchange rate suffered an extreme negative trend in the 1980s: it fell by 49 percent between 1980 and 1989 (table 5). The data available for manufacturing exports, through 1988, also show Peru as the only country included in table 4 for which manufacturing exports decreased. The fall in the real exchange rate was concentrated in the period beginning in 1985, under a government which controlled the exchange rate (or more exactly a set of differential rates) and used that control to hold down the real exchange rate in a vain effort to stop inflation. The new government in 1990 disavowed such controls and adopted a floating exchange rate as one aspect of its adjustment program. The government recognized that the real rate had been driven much too low under controls and expected that freeing the rate would lead to a rapid depreciation, favoring reactivation of exports. But the free market approach did not work as expected, nor has it worked in the year following: market forces have persistently held the price of foreign exchange far below levels necessary to restore export incentives.[12]

The main reason that the floating exchange rate failed to float in the direction needed to correct overvaluation was the intense liquidity squeeze imposed by the new Peruvian government in its determined effort to stop inflation. The squeeze on liquidity drove interest rates so high that asset holders switched from dollars to domestic currency on a large scale, driving the real exchange rate

---

[12]See Sheahan n.d.: chap. 3, for a country study of Peru being prepared by the Federal Research Division of the Library of Congress.

down instead of up. An additional factor specific to Peru was that dollars earned from drug exports also fed into the market and helped hold down the price of dollars. The net result added to the forces acting to restrain inflation but at the same time blocked the possibility of gaining room for recovery through increased legal exports. A full year after the adoption of the adjustment program the real rate still remained far too low to encourage export growth.

Given the current emphasis of the Mexican government on encouraging direct foreign investment, and the growing positive response of foreign investors, there is some danger that a market-determined real exchange rate will be too low for continued growth of industrial exports or for the ability of Mexican industry to compete against liberalized imports. This is a critical point of tension between market-based economic policy and structural concerns. If industrial growth is in fact among the nation's goals, it might help to place limits on the inflow of foreign capital and to set targets for the real exchange rate as supporting conditions for industrial development, even though such actions would constitute partial repudiation of reliance on market forces. They need not imply repudiation of a more open economy than in the past, with greatly reduced trade restrictions and much more emphasis on competitive exports; they need only mean an assertion of responsibility for the direction the Mexican economy is to take in the future.

Liberalization and the Structures of Trade and Production

Latin America gained some advantages from the strategy of import substitution but suffered high costs from carrying it too far, with inadequate attention to its dangers. It would be similarly likely to gain some advantages but to suffer serious losses from any all-out reversal to liberalization. Unhappy results with one kind of exaggeration rarely mean that the opposite will be any better. Latin American economic systems may well become more efficient and more capable of sustained economic growth under outward-oriented trade regimes, but these goals need not require countries to allow complete freedom for foreign investment or capital movements in general, to follow the existing pattern of comparative advantage in all decisions on production and investment, to retreat from the use of taxation and social programs to moderate inequality and poverty, or to repudiate the responsibility of the state to help shape the character of the nation's development.

Mexico's steps toward trade liberalization in the mid-1980s offered real promise of gains, both for adjustment to the debt crisis and for long-term growth. Its more dramatic moves under the Salinas administration toward full liberalization of both trade and investment, away from internal intervention and previous social programs, could be more troublesome. But much depends on exactly how it is done, with what kinds of protective offsets and what goals. The question is not so much whether liberalization is or is not desirable but what it means and how it is implemented.

Trade liberalization could be consistent either with continued and possibly strengthened industrial growth or with a setback to industrialization that might take a long period to repair. Whatever it does to the overall level of industrialization, it is bound to change the relative profitability of investment among industries, setting some back and stimulating others. The changes would surely be in part determined by the response capacities of particular firms, but they might also have systematic patterns. They might lead either toward or away from more labor-intensive activities, with important implications for employment opportunities and income distribution. They might lead either toward or away from industries using more advanced technologies with greater learning opportunities.

The main ground for concern about possible contraction of the industrial sector is the fact that it has been built up by many years of protection, under which firms could prosper despite high costs or poor product quality because domestic buyers could be denied access to superior alternatives. Much of Mexican industry earned its profits chiefly from taxes on customers, in the form of prices above the levels of potential imports. Protection from import competition was often supported by a mutually agreed absence of domestic competition, given acceptance of monopolistic behavior by governments eager to stay on good terms with existing private business. With nothing resembling any antitrust policy internally, with cultural presumptions in some respects unfavorable for competitive behavior, and with protection from potentially competing imports, Mexican industry could usually count on survival almost regardless of efficiency or product quality.

To sweep away protection from import competition could wipe out many such firms. Some of them are surely capable of raising efficiency and product quality enough to survive and even grow in the face of open competition from imports, but some are bound to contract or to be driven out of business completely. Sustained industrial growth is not likely unless contraction on the side of previously protected industry is offset by expansion of industries able to compete both against imports and in external markets. Absent an adequate offset on the side of exports, the industrial sector as a whole clearly could be driven down, forcing employment away from industry toward primary activities and services.

A wholehearted liberalization strategy, allowing the structure of production and trade to be determined by comparative advantage, could favor industrial growth if comparative advantage is on the side of industry, or it could set it back if not. A strong natural resource base, providing good export earnings from natural resources while holding down requirements for imports of raw materials, works against industrialization by keeping comparative advantage on the side of primary production. In such a structural context, industry has to fight uphill against market-determined incentives. The dramatic postwar successes of the East Asian countries in achieving fast industrial growth through outwardly oriented economic strategies have in a sense been greatly aided by the absence of abundant natural resources: that absence directed their comparative advantages to the side of industry. Industrial exports were naturally based in the early stages on low-cost labor and subsequently, as successful growth made labor scarce and more expensive, on more technologically advanced industries. Orientation toward industrial exports was for these countries perfectly consistent with relative costs and incentives in competitive markets. Liberalization in Latin America, with much stronger advantages in terms of natural resources, need not have the same effect at all because market forces may not favor industry to anything like the same degree.

The possibility of a setback to industrialization is exemplified by the first decade of liberalization in Chile. Investment proved to be exceptionally weak relative to prior historical experience, and industrial output as of 1984 remained lower than in either 1970 or 1974 (*Colección Estudios CIEPLAN* 16 [June 1985]). Many industrial firms were wiped out, particularly in textiles and other consumer goods previously sheltered by high rates of protection, without a fully compensating rise of new industrial activities. That setback to industrial production and employment was not a necessary result of liberalization itself, but it demonstrated what can happen if the process is mismanaged.

The two most powerful negative factors in Chile's case were a prolonged policy of extremely high real interest rates, discouraging investment in productive activities generally, and the recourse to a fixed exchange rate in the latter 1970s as a means to try to stop inflation (Ramos 1986). Liberalization did not require these choices, but it made their results more costly than they otherwise would have been. Overvaluation of the currency hurts exports in any context, but if accompanied by import liberalization it can distort the production structure by destroying even efficient firms that could have survived if they had protection to offset the overvaluation. Chile had to go through one of Latin America's deepest depressions, from 1981 through 1983, before its macroeconomic policies were corrected. The correction process also involved a moderate increase in protection, without reversing the basic policy of a highly open economy. That new combination proved much more successful, supporting renewed growth of industry along with the rest of the economy. The traumatic experience up to 1984 was not a defeat for liberalization but a reminder that the results can go either way, depending on the whole context of economic policy.

There is no reason that Mexico need waste a decade or more of weak industrial growth under trade liberalization, but it could do so if the real exchange rate is too low for industry to compete successfully. The key test is that discussed in the preceding section: the profitability and growth of industrial exports. They grew so successfully in the course of the 1980s that the *possibility* of continued industrial growth under liberalization is abundantly clear. But the favorable real exchange rates essential for that success in the 1980s reflected two factors which could conceivably be very different in the future: (1) weak markets for oil exports which restricted foreign exchange earnings and thereby kept up pressure on the government to maintain a high price of foreign exchange in real terms, and (2) the absence of capital inflows that could have held down the price of foreign exchange by increasing its supply.

A resurgence of earnings from oil exports or major growth of other primary exports, or of tourism and other service exports, would have a two-sided effect on the structures of production and trade. By increasing both national income and the supply of foreign exchange it could stimulate aggregate growth, but at the same time it could pull investment and production toward the growing sector at the expense of industrialization. All else equal, it would make industrial imports cheaper and industrial exports less profitable. That is to say, it could change comparative advantage against industry, against the pattern of growth that has been so successful in East Asia.

Market forces might provide an answer to any such adverse turn for the industrial sector through more rapid growth of national income. Rising national income could stimulate industrialization both by providing stronger domestic market opportunities and by pulling up the price of foreign exchange through increased demand for imports. Absent deliberate intervention, the balance could go either way. If the balance threatened to go the wrong way, if a fully open system began to squeeze out industry from the side of comparative costs, that effect might be offset either through differential exchange rates or sterilization of export earnings. Public ownership of the oil industry provides a control mechanism that would be absent under private ownership. The Mexican government does not need to buy from private exporters the foreign exchange earned by oil; it can in principle accumulate the exchange as additional reserves without increasing the money supply (or, alternatively, use a greater share of the earnings to reduce external debts). That would mean holding domestic resource use below national income, as the de la Madrid administration did under pressure from external debts, but now more as a matter of longer-term strategy to keep up incentives for industrial exports.

The other temporary factor favorable for real exchange rates and for industrial exports in the 1980s, the absence of net capital inflows, was hardly a policy choice in that period but may become so in the 1990s. It becomes a policy choice when both domestic and external investors regain confidence in the Mexican economy and try to share in its gains by moving capital into the country. That process seemed well under way in 1990. The question is what level of net capital inflow is actually desirable, if any. With complete liberalization in the sense of a hands-off attitude toward capital flows, the question makes little sense. Private capital will move in if the returns look promising, and will not if they do not, to degrees determined by private decisions. But such private decisions can raise severe macroeconomic problems. Capital can move out so rapidly that it cripples ability to finance necessary imports or, on the contrary, it can move in on such a scale that it creates pressure for a real exchange rate that is too low to allow domestic industry to compete successfully. If the latter possibility is seen as a problem — if industrial growth is sought for its own sake as a desired component of long-term development — then the government may need to establish something like a desired target, or limit, for net capital inflows.

Interviews with Mexican economists in early 1991 consistently suggested a consensus on the desirability of renewed net capital inflows as a key component of liberalization. In this view, it is both normal historically and efficient in terms of resource allocation to allow capital to flow from countries where it is more abundant to developing countries in which it is relatively scarce. David Ricardo would approve of the argument, but a problem remains: in an economy characterized by strong natural resource availabilities and by a liberalized economic system, high capital inflows can make foreign exchange too cheap for domestic industry to thrive. Does that matter? Is the goal essentially one of external financial balance regardless of output structure, or is it instead one of fostering industrialization in the process of development? If the latter, the government needs to have a target for an upper limit to capital inflows even if — or more exactly, especially if — the outside world wants to move capital into the country on a large scale.

Assuming a combination of market forces and intervention that yields an exchange rate favorable for continued growth of industrial exports, along with rising industrial imports following the pattern of comparative costs, what kinds of industries are likely to thrive and what kinds to either collapse or be severely constrained? The traditional Heckscher-Ohlin theory of factor proportions would suggest that labor-intensive activities would gain and those that are capital intensive, or need to be able to keep up with rapid technological change, would lose. If that proved to be the actual pattern, liberalization combined with growth of industrial exports should be favorable for employment, wages, and equality. By the same token, it would be adverse for learning and technological change, shaping the industrial sector toward traditional industries with established technologies at the cost of lower participation in those with most rapid technological advance.

Empirical studies of Mexican trade patterns do not support the presumptions of the Heckscher-Ohlin analysis at all well. Neither Mexican industry in general nor industrial exports have in the past been strongly labor intensive. Growth of industrial employment has been slow relative to growth of industrial production, and manufactured exports have been even less labor intensive than Mexican industry in general (Behar 1988: chaps. 6, 7, and 12). That historical picture might be explained in two very different ways, with contrary implications for consequences of liberalization.

The more traditional explanation is that intervention by protection plus differential export subsidies, accompanied by frequent overvaluation of the currency, was biased in favor of capital-intensive activities. Protection made it possible for industries using scarce and expensive inputs to be as profitable as industries that took greater advantage of low labor costs: intervention divorced profitability from efficiency. Export subsidies that took the form of differentially low-cost credit, tax favors based on capital investment, and overvaluation holding down prices of imported capital equipment, all stacked the cards in favor of capital- rather than labor-intensive production. To the extent that liberalization does away with such bias, substituting an export-oriented exchange rate for subsidies and protection, it should raise the labor intensity of both exports and domestic production.

The alternative explanation, based on changes in trade theory in recent years, has very different implications. The changes in trade theory apply specifically to trade in manufactures and respond to a growing conviction that Heckscher-Ohlin analysis in terms of relative factor supplies and costs does not account for the main lines of postwar trade patterns (see especially Krugman 1980; Helpman and Krugman 1985). A rising share of trade in manufactures consists of exports and imports going both ways between the same industrial branches, using similar factor input combinations. Trade among the industrialized countries is less and less a matter of differences in factor proportions and more intra-industry in character, based on scale economies, ownership relations, leadership in technological change, and behavioral characteristics of particular firms. Liberalization of this kind of trade does not lead to concentration of whole industries in particular countries but to greater focus and larger scale in specific product lines within industries that may be growing simultaneously in all the trading countries.

Such two-way exchange within industrial groups contrasts strongly with the more resource-specific exports and imports characteristic of much of the trade between developing and industrialized nations. The Heckscher-Ohlin interpretation fits the trade of developing countries much better than it fits trade among the industrialized countries. Where does Mexico belong in this dichotomy? In many respects it is still a developing country, but in its exports of manufactures it is now more accurately characterized as a newly industrializing country, very much involved in the intra-industry type of trade common among the industrialized nations. A classification of Mexican industries in terms of the degree to which their combined exports and imports consist of the industry's own products identified 34 of 184 industries as characterized predominantly by intra-industry trade: they accounted for 23 percent of industrial value added in 1980 but fully 51 percent of all industrial exports for the period 1978–83. Compared to the rest of Mexican manufacturing, these 34 industries are disproportionately involved in both exports and imports, are distinctly more capital intensive, and are characterized by high degrees of economies of scale (Ros 1991, especially tables 2 and 4). Liberalization is helping some firms within these activities by reducing the costs of their inputs and hurting others by driving down the prices at which their products can be sold in domestic markets, but it is not acting strongly to wipe out whole branches of any industry or to change greatly the structure of demand for domestic factors of production.

This newer type of explanation of trade in manufactures seems much more relevant than the first as a guide to likely outcomes for Mexico under liberalization. It does not offer high promise of gains for employment, except insofar as a more dynamic process of technological change favors faster growth of the economy as a whole. The outcome might be more favorable for learning and for close connection to technological change, though that is a complicated question. These exports are of many different types, some favorable for learning and technical change and others much less so. And within the varied set, those exports most clearly identified with high rates of technological change are also those most closely associated with foreign rather than national ownership.

Kurt Unger distinguishes four main categories of Mexican industrial exports: (1) products based mainly on natural resource advantages, such as petrochemicals and processed foods; (2) those based on advanced technology, including automobile engines and office machinery; (3) products of mature industries with relatively stable technology, including basic chemicals, glass, and plastics; and (4) standardized parts for the automobile and other industries, for which the key to successful competition is more a matter of economies of scale than of advanced technology. The most dynamic growth in the decade 1975–1985 was in category 2, with the most advanced technology but also with predominant ownership by transnational firms. National firms are best represented in categories 1 and 3, with considerably lower opportunities for learning (Unger 1990: 59–60, 93–97, and 191–94).

With the near-removal of restrictions on foreign investment in the last few years, ownership of firms involved in exports may well change considerably. It is always possible for Mexican firms to reach out and buy foreign companies, as the leading Mexican producer in the glass industry did recently. But the more likely expectation for the near future, especially for the more technologically dynamic industries, is that foreign ownership shares will increase. If the emphasis of foreign firms continues to be on the industries with relatively advanced technology, the learning component of Mexican industrial exports may not be widely diffused within Mexico. It even becomes a question of whether it makes any sense to term such exports "Mexican" if the firm-specific technology is controlled by foreign firms. From the perspective of joint gains of efficiency and of knowledge in a more integrated market between Mexico and the United States, the question may not matter greatly. It matters more if one places differential value on gains of knowledge and technological capacity by Mexicans. If one does, unrestricted foreign investment under existing conditions can take away much of the meaning of likely gains in knowledge from industrial exports.

The main factors under existing conditions that work against Mexican participation in more technologically dynamic industries are (1) the existing distribution of such knowledge between Mexico and the more industrialized countries, giving the latter a strong comparative advantage in these fields, and (2) the low quality and restricted access of the Mexican educational system, wasting the potential of a high proportion of the country's population. The most important question for the future in this respect, whether or not foreign investment remains unrestricted, is the degree of determination and success of the society's efforts to make educational opportunities more nearly equal, to broaden access to skills and capacity to respond to change. Success in this key ingredient of development could improve the prospects for lessened inequality in many ways and could at the same time raise Mexican capacity to compete in fields of rapid technological change.

The gains in both employment and learning from export competition itself could also be broadened by specific measures to help small and medium-sized firms participate more fully in the export process. These firms are relatively labor intensive and are almost entirely under Mexican national ownership. In addition, greater growth opportunities for them could help increase the flexibility of the economy by reducing the degree of ownership concentration and increasing the degree of domestic competition. Few of these firms were exporters before the changed policy regime of the 1980s, but they began to move this way during the decade.

Analysis of a sample of small and medium-sized industrial firms in 1990 shows that 10 percent of them had become direct exporters and that two-thirds of this group had entered export markets for the first time in the 1980s (Ruiz Durán 1991). Still, the most promising role for small and medium-sized firms in connection with exports may not be through direct exports themselves but as subcontractors and parts producers working with larger firms. As this study emphasizes, their flexibility often gives them a positive base for specialized subcontracting, but active help from the government to encourage subcontracting could make a significant difference. It is also true that exchange rate policies will matter greatly: subcontracting is most unlikely to make any great headway if imported parts and supplies are systematically less expensive than domestic alternatives.

Sustained encouragement of industrial exports should favor opportunities for learning through participation in external markets and enable firms to enter progressively more dynamic fields. But how much Mexicans themselves can respond depends on what is done to foster widespread access to good quality education and skills, to encourage research and development, and to create favorable conditions for entry and growth of new firms. Given positive state leadership in such directions, liberalization and increasing trade in manufactures should permit successful competition in increasingly advanced fields; absent such leadership, they will not.

Liberalization, Wage Policies, Poverty, and Income Distribution

Liberalization in the restricted sense of reducing trade barriers and encouraging exports could be managed in ways likely to improve employment opportunities and lessen poverty. Liberalization in the more general sense of allowing market forces to work freely, without regulatory intervention or subsidies, would involve some changes that might act to lessen prior degrees of inequality and poverty and others more likely to worsen them. In Mexico's case it would mean a movement away from a complex system that gave advantages to many of the rich but also support for many of the poor, to a system in which incomes would be more nearly determined by relative scarcities and demand, in a country where capital and skills are relatively scarce and ill-educated labor is abundant.

Mexico's economic strategies since the Revolution have gone through many different phases but have consistently maintained an inclusionary kind of corporatism.[13] That has meant much more than protection and selective favors for industry. It has meant both support for and control of wages, with gradually rising minimum wages that directly affected earnings of a majority of the urban labor force, with social security and public health services, institutionalized channels for negotiation with both peasants and workers, extensive subsidies for basic foods, and in some periods support prices for the output of peasant producers. The system was shot through with inefficiencies and special privileges and was constantly changed in operating details, but it held the society together in ways that a fully liberalized system might not have been able to achieve. During the first years of contraction in response to the debt crisis, preceding the shift in 1985 toward liberalization, it helped considerably to moderate the impact of

---

[13] Bazdresch and Levy (1990) emphasize the negative consequences of Mexico's protective society but bring out the positive goals as well.

aggregate contraction on the poor (Lustig 1989; see also Austen and Esteva 1987). Interferences with markets and prices always have costs for efficiency but they can have genuine advantages for the weak.

Mexico's system of intervention and control did not succeed in preventing an exceptionally high degree of inequality, compared either to the northern industrialized countries or to most developing countries. Within Latin America, Mexico's distribution of income at the beginning of the 1970s was not greatly dissimilar to those of Brazil, Colombia, or Peru, but it was much more unequal than in Argentina, Chile, Costa Rica, or Uruguay (see Ahluwalia 1974; Ahluwalia, Carter, and Chenery 1979). Whether Mexico's diffused kind of intervention lessened or worsened inequality on balance is uncertain, but given the power of the underlying factors making for inequality it may well be that the system accomplished a great deal simply by keeping the degree from getting even higher.

The factors generating high inequality in Mexico's past can be divided into two groups. One group is a powerful set of causes operating through market forces, making the price system systematically adverse for equality. The second is a set of causes that derive on the contrary from particular kinds of intervention acting to block or distort markets. Just because markets aggravate inequality in many ways does not mean that all kinds of intervention are helpful. Some kinds of intervention provide help for the poor precisely because they limit the operation of market forces, but some on the contrary make things worse than they would have been in the absence of the interventions.

Market forces have in many respects worked to aggravate inequality because of the abundance and high rate of growth of unskilled labor relative to the growth of opportunities for productive employment, a weak system of public education unable to provide anything like equal opportunities, relative scarcity of the skills able to command high wages, a high concentration of property ownership, and a fiscal system in which taxation of business income—or personal incomes of property owners—has long been kept exceptionally low (Sheahan 1987: chaps. 2, 3, and 11). These are not primarily problems caused by state intervention in the economy. Liberalization might be helpful with respect to some of them but might at the same time make others more negative than they have been in the past.

While market forces themselves have been in many respects unhelpful, so have some particular forms of intervention. Two major examples have been the use of high protection for industry under the strategy of import substitution, and of price controls on basic foods when applied at the level of low-income agricultural producers. A third possible example, or perhaps more a question, concerns the distinctive Mexican system of land use, the ejidos created in the great land reforms of the 1930s. The small producers were given rights to use the land and to pass on these rights to their children, but not to sell it. The point of forbidding sale was to prevent a possible return to high degrees of ownership concentration and to keep the small producers on the land. Given the restrictions on their right to sell their land and to mortgage it, the small producers have not had full access to normal commercial credit. The state has therefore provided public credit, on favored terms but subject to conditions that limit choices of producers to some degree. Their incentives under this system have been directed toward producing basic foods for themselves or the domestic market, even if alternative crops have higher market value. The system thus tied up choices of rural small producers, who make up a high proportion of Mexico's poor (Levy 1991). All three of these kinds of intervention have had negative effects on income distribution; the question is whether or not eliminating them is the best way to correct the problems they have raised.

If the current process of trade liberalization comes close to eliminating the first of these three kinds of intervention, the rural poor and workers in the urban informal sector should gain. Protection of industry acted to favor profits of the industrial sector at the expense of the general public, to raise prices of industrial products relative to income of rural labor, and to lessen the labor intensity of industrial production. That is true of developing countries in general, at least as suggested by cross-section tests of the relationships between protection and income distribution (Bourguinon and Morrison 1989). The strategy of import substitution considered separately (as distinct from the whole system of inclusionary corporatism) probably contributed to inequality in Mexico. A movement away from it could be helpful. But the likelihood that it will be helpful depends on how it is done and with what accompanying policies.

The contrasting experiences of Chile and Colombia when turning to more outward orientations help illuminate some of the choices and possibilities facing Mexico. In the case of Chile liberalization was associated with worsening inequality and deepening poverty for about fifteen years. In Colombia the more circumscribed turn outward proved to be helpful in reducing poverty but still drove down industrial wages for a decade: the changes in income distribution were initially adverse for workers in the middle of the distribution while favoring both the poor and property owners. The subsequent trend through the 1980s continued to be favorable for unskilled workers and gradually pulled real wages in industry back up as well. With the help of strong demand conditions and rising employment, sustainable even through the 1980s thanks to avoidance of heavy borrowing prior to the regional debt crisis, Colombia's high degree of inequality began to come down for the first time in the postwar period, and continued steadily downward through the mid-1980s.[14]

Three major differences between the outward turns in Chile and in Colombia help account for heightened inequality in the former case and gradually lessened inequality in the latter. The first major difference is that Chile's liberalization in the 1970s was an all-out and immediate affair, while Colombia's transition from 1967 was much more gradual and carefully incomplete. The second is that Chilean macroeconomic policy stressed monetary and fiscal restraint to eliminate inflation and fell into the trap of using the exchange rate to eliminate inflation rather than promote exports, while Colombian macroeconomic policy aimed instead at raising employment and industrial exports through an exchange rate policy guided by the consistent objective of promoting such exports. The third is that Chilean liberalization was poisoned by its conception and execution as an ideological goal, a determination to get away from social programs as well as economic regulation, while Colombia's turn away from protection and toward export promotion was linked to a program of social action intended to lessen poverty and inequality as a complement to increased concern for efficiency. Both countries turned more toward reliance on market forces, but the Chilean government did it with righteous conviction and the Colombian with great circumspection.

All-out liberalization in the Chilean style means that protected industries have little time to carry out possible adjustments before they go under, and they are likely to go under more quickly than the competitive side of the industrial sector can expand into export markets. The immediate consequence in Chile was higher unemployment and wider damage to investment prospects than a gradual process need have involved. That negative context was compounded by tight monetary and fiscal policy, at first partly compensated by stimulation of new exports through the exchange rate but then worsened by freezing the exchange rate in order to curb inflation. The rise in poverty in Chile during this period should not be blamed on trade liberalization itself, but it is testimony to the damage that can be caused by rapid liberalization in the absence of positive intervention to lessen the negative impacts of market forces on the poor.

---

[14]On Chile in the 1970s, see Cortázar 1980. On Colombia, see Urrutia 1985; World Bank 1990: 43, 48–49.

Colombia's much more gradual shift away from protection and toward export promotion avoided any drastic cutback in existing industries, without preventing an important redirection of incentives. Moderate reduction of existing protection and a new resistance to requests for high protection served to put brakes on wasteful investment, while adoption of a crawling-peg exchange rate geared to promotion of industrial exports offered a new promise of profitability to firms capable of competing in external markets (Urrutia 1981). Monetary and fiscal policy, and institutional changes favorable for private investment, were used to raise aggregate demand and production even at the risk (and with the unintended consequence) of increased inflation. At the same time, social programs, including land reform, were used to attack poverty, in the conviction that social action is in no sense contrary to the success of a more market-oriented economic system.

These accompanying policies in Colombia succeeded in reducing poverty as the reorientation of the productive system began, in strong contrast to the Chilean result. Real wages of unorganized workers at the bottom of the income scale—rural labor and construction workers and even household maids—rose impressively while external trade was opening up. The proportion of the population below a fixed poverty line fell from 41 percent in 1971 to 25 percent by 1988, while the Gini coefficient of inequality fell from 0.54 in 1970 to 0.50 by 1976, and to 0.45 by 1985 (World Bank 1990: 41, 49).

While it proved to be possible to reduce both poverty and inequality in Colombia in the process of redirecting industry toward export competition, it may still be argued that this redirection had serious negative effects. Along with the highly desired objective of making industry more cost conscious and more concerned with productivity, it made industrialists bear down more forcefully against wage increases, while the government itself backed away from wage support. Real wages of industrial workers were driven down from 1967 through 1972, and the slow rise that set in thereafter still left them below the 1970 level as late as 1980. Although unorganized workers with lower incomes gained relative to organized labor, property income rose relative to wage earnings for a long period.

On another level, the new emphasis on industrial exports may have contributed to an early demise of land reform. Possibly because of a new preference to encourage rural labor to move into urban employment, or possibly because of violence associated with the process of land reform, the leaderships of Colombia's two dominant political parties agreed in 1973 to abandon the program

(Zamosc 1986). That points to a process that could be important for Mexico: liberalization in the sense of withdrawal from intervention in the economic system has a dynamic of its own, reinforcing that side of policy preferences adverse to any limits on private property rights, and to any social program that may impinge on them, regardless of any necessary connection to removal of regulatory barriers.

The second example of previous intervention adverse for the poor was the use of price controls on basic foods, applied at the level of the producers. This long-established form of food policy in Latin America happily seems to be losing ground everywhere: it was intended to benefit the urban poor but it helped many of the urban nonpoor as well, hurt the rural producers—with a much higher percentage of families in poverty than the urban sector— and discouraged output of basic foods. Liberalization in the sense of eliminating this kind of price control is almost surely an improvement from the point of view of reducing poverty and inequality. But there is no need to stop there if the goal is really to help the poor. Mexico adopted a potentially preferable strategy in the 1970s, paying subsidies to producers of basic foods and at the same time subsidizing their prices in lower-income urban neighborhoods. The main problem with that approach was that the subsidies were not covered by adequate taxation and therefore contributed to excess demand, inflation, and the growth of external debt. The whole program was cut back severely in the fiscal austerity imposed by the debt crisis. Full liberalization of the economic system would rule it out in the future. The alternative of imposing more effective taxation on higher-income people in order to restore the subsidies without such inflationary consequences would avoid the negative consequences of the past with a much more favorable result for the poor.

The third example of possibly harmful intervention, the ejido system placing many small rural producers partly outside the market system, raises issues on several different levels. The proposal to do away with it strikes at a distinctive institution long believed to be at the core of Mexico's inclusionary corporatism (see Collier 1981). It is possible that the system has played a key role in holding the society together in times of stress, to a degree often envied in the rest of Latin America. It is also possible, though not certain, that doing away with it now could improve the position of the rural poor. Santiago Levy's study of poverty alleviation gives good reasons for expecting a positive result but becomes more doubtful in its position against any related interference with private markets, especially in its more general conclusion: "Primary poverty in Mexico is the result of institutional structures and interventions in resource allocation mechanisms which undervalue the resources owned by the poor" (Levy 1991: 81).

Eliminating restrictions on land sales under the ejido system, and on crops eligible for support by public credit, could be helpful for that fraction of the rural poor that would gain full title to the land they use, but not nearly as helpful for the rural poor as a whole as these moves could be if they are accompanied by alternative forms of intervention. More helpful kinds of intervention could include price supports for production of basic foods, tax-financed subsidies as suggested above, and limits on the size of permissible landholdings to prevent a return to high degrees of concentration. Larger average landholdings are often associated with less labor-intensive kinds of crops and methods of production: in the context of surplus labor, with inadequate opportunities to move into productive employment outside of agriculture, changes in this direction could hurt landless rural labor to a degree outweighing any gains to the small landholders who sell their land. All-out liberalization argues against restrictions on the size of landholdings as against any other intervention in agricultural markets. All-out liberalization is not optimal for the poor.

Intervention in labor markets to support wages has similarly had multiple effects, with mixed gains and losses in social terms. It has served to give Mexico a degree of labor peace rare in Latin America and at the same time, by the same token, to hold labor in a relatively dependent position. It has in the past served to limit downward pressure on wages in conditions of weak employment opportunities. It has helped hold up labor's share of national income and at the same time favored higher-income organized workers relative to the self-employed and the rural poor. The separable component of protection for minimum wages has had a more egalitarian effect than support for average wages in manufacturing, though it too had the drawback of favoring urban workers over rural labor.

Both types of support for labor seem to have gone into reverse in the 1980s, with government intervention bending more toward wage restraint than toward protection. The fall in average real wages in manufacturing is explicable readily enough in terms of weak labor market conditions, but to allow such a large decrease just because of weak labor markets is a significant change. The goal of a more competitive industrial sector, in the context of liberalization, must have been a major factor determining this change. In that sense, liberalization contributed to the shift of income distribution adverse to labor. Still, continued trade liberalization need not hold real wages down in the future. As discussed in the preceding section, real wages can rise if labor productivity rises, provided that macroeconomic management can, by avoiding periods of overvaluation of the real exchange rate, also avoid the need for further drastic devaluation in real terms.

The especially deep cut in real minimum wages in the 1980s may also be explicable in terms of the logic of liberalization, placing increased emphasis on efficiency principles. The question of support for minimum wages is always a crucial dividing line between concern for maximum economic efficiency and concern for poverty. The move toward liberalization seems to have pushed the balance in Mexico further toward the efficiency side, at the expense of real incomes for those at the lowest wage levels.

The downward thrust of weakened employment conditions and falling wages was offset to some degree by extensive social programs in the first half of the 1980s, but then the drive toward liberalization, combined with fiscal austerity, began to cut them back greatly as well. Some reduction in social spending was essential to curb the excesses of the López Portillo government, but the extreme degree chosen may have been a consequence of the logic of liberalization, calling for scaling down the role of the public sector. A fiscally responsible alternative might have been to raise taxation on property incomes to finance greater efforts to lessen the costs to the poor, but the actual choice during the macroeconomic contraction was to concentrate on cutting out the programs.

As resource outflows stopped in 1989–90 and evidence of beginning recovery became clearer, the Salinas administration began a new set of social programs with a considerably different orientation. The new approach, the "National Solidarity Program," or PRONASOL, emphatically rejects the broad system of subsidies and high level of social expenditure of the late 1970s, in favor of a much more selective approach.[15] In principle, the emphasis on selectivity could help raise efficiency in targeting help for the poor. This is the counterpart in the social field of the efficiency-oriented logic of liberalization. It could cut down on social expenditures that mainly benefit the nonpoor and could help direct public support toward improving the productive capacities as well as the lives of the poor. Or it could become a token offset to much more powerful forces working to increase inequality; a public-relations rationale intended mainly to help keep the governing party in power and to defend a basic economic strategy that changes the whole social balance adversely.

Analysis of the National Solidarity Program by Denise Dresser brings out both genuine gains for particular communities of the

---

[15] Villarreal (1990) emphasizes both the new selectivity and the role of Mexico's social policies as an essential complement to increased emphasis on efficiency.

poor and serious problems for PRONASOL's relationships to the possibilities for more meaningful democracy (Dresser 1991). In some respects the program has created new openings for initiative by local groups to take their own lead in self-directed activities and for participation in the program by opposition groups. In other respects it has been an additional channel for patronage tying people to the PRI and a means to target voters in areas that went against the party in the 1988 presidential election. It constitutes a powerful weapon of partisan advantage, used to fortify the dominant party's control of the state. Its financing and its choice of projects have been highly "personalistic." "There are no democratic means of control over the PRI's spending of PRONASOL funds or over the individuals and organizations that carry out Solidarity's diverse programs" (Dresser 1991).

The appeal of the National Solidarity Program is that it could help many of the poor, but it may at the same time preclude meaningful electoral competition to the PRI. The conflict is familiar in many countries: political parties can act to help the poor in ways that mainly help themselves. But that is not necessarily adverse for democracy; it is one of the ways that democracies work. Where the existing system provides careful legislative review of expenditures and methods, where a critical press searches for signs of abuse, and where aggressive political distortion can become a handicap in following elections, the chances of getting social programs that help the poor without blocking political competition can be fairly high. The basic political problem with the National Solidarity Program is not inherent in the program itself: it is the absence, so far, of the constraints of genuine democracy.

Orientation and Character of the Mexican State

Changes in economic strategy interact with political and social factors in ways that are too complex for any definitive analysis in terms of economic forces, or of any other single-discipline approach for that matter. Still, it should be possible to clarify some of the ways in which a fundamental change from a protected corporative orientation to an open economy mainly shaped by market forces is likely to affect the balance of pressures and preferences within the society. Some of them may work in the direction of a more equitable and open society, and some may on the contrary generate strains adverse to both equity and democracy. The issue is not so much one of trying to decide between two opposing generalizations—that economic liberalization will promote or will work against equity and democracy—but rather trying to identify key strands of interaction.

The first of the connections considered below is how the consequences of liberalization for the distribution of income may affect the willingness of a majority of the people to accept such an economic system, once given any real political choice, and thereby in turn affect the willingness of the government and of those who gain most from such an economic system to accept majority decisions in elections. The main hypothesis for Mexico, as for much of Latin America, is that reliance on unguided market forces is likely to aggravate inequality and thereby tensions between economic and political liberalization.

The second and directly related connection is how an increased emphasis on private investment and private control of production decisions, with greatly reduced direct intervention or state investment, may restrict the government's range of feasible choices on such issues as tax policy, public expenditures, and social programs. The hypothesis is that it means a loss of power to implement any democratically chosen course of action running contrary to market forces. The government loses power to respond to public preferences. In a sense, that is the essence of the case for liberalization: that market forces rather than political decisions will determine how the economy works. The question is the compatibility of this course, under the structural conditions of Latin American economies, with a democratic political system.

The third set of connections concerns the ways in which the historical context of Mexico during its transition toward economic liberalization differs from those of prior liberalization movements in Brazil in 1964 and the Southern Cone in the 1970s. The hypothesis is that the striking contrast between Mexico's relatively nonviolent process and the extreme repression that characterized the earlier experiences is mainly explained by its adoption under a well-established corporatist system. It did not start from a democratic context, but neither was it associated with violent overturn of prior democracy. An accompanying hypothesis is that this corporatist system itself, while facilitating the economic transformation, runs contrary to any complete democracy.

The fourth and final connection considered below goes from the political side to the economic. What does it mean for the character of economic liberalization if political liberalization is delayed indefinitely? Is it asking too much for Mexico, or any other previously restricted society, to adopt both kinds of liberalization at the same time? The main hypothesis in this connection is that delayed political liberalization would not prevent sustained economic liberalization, but it would allow it to take forms making for higher degrees of inequality and more strongly fortified positions of privilege than would be likely if the process of economic liberalization were constrained by effective expression of public preferences.

MARKETS, INEQUALITY, AND DEMOCRACY

The main reason for doubt about the compatibility of liberalization and democracy under the structural conditions of the Mexican economy is that the new economic strategy could aggravate inequality. If it persistently holds down incomes of unskilled workers relative to those of the minority with scarce skills or property ownership, a democratic political system would be most unlikely to allow it to remain the nation's basic economic strategy. If most people lose by it, in either absolute or relative terms, that should make it readily possible in a democracy to attack and change the economic strategy. Such a resolution could maintain democracy by curbing liberalization. But it could also turn into a threat to democracy from the side of the society determined to adopt or to maintain liberalization, either because of conviction that it is essential for successful development or because it works in their interest.

The structural conditions of the Mexican economy have in important respects made market forces work in the direction of heightening inequality. The main factors discussed above have been an abundance of unskilled labor, unequal access to education and skills, and highly concentrated wealth. Despite these adverse factors, common to most of Latin America, some of Mexico's structural conditions are less adverse for equality than those of the poorer Latin American countries. In particular, reduction of land ownership concentration by the reforms of the 1930s and, subsequently, gradual progress in opening access to education and skills, state ownership of the nation's oil resources, and the remarkable success of the 1980s in developing competitive industrial exports that offer growth of employment opportunities in a new direction, all improve the potential for lessening inequality. Mexico is between the worst of Latin America's adverse conditions for combining equity with efficiency and the opposite kinds of positive conditions characteristic of Western Europe. The balance is too doubtful to make it safe to rely on market forces but not so inexorably adverse as to prevent successful combination of equity and efficiency if selective intervention is used to shape market forces to that end. The longer-term result depends on exactly what is done to change structural conditions in ways favorable for lessening inequality.

The steep declines in wages accompanying Mexico's moves toward liberalization in the 1980s, and the rise of property income relative to labor income, were consistent with the expectation that the process of liberalization implies increasing inequality, at least until the adverse underlying conditions of the economic structure can be changed. Decreasing wages were in part a necessary component of macroeconomic adjustment to correct aggregate excess demand in the first place, but the degrees to which both average and minimum wages were cut, especially the latter, may well have been a result of the logic of liberalization, favoring both withdrawal from wage support and a goal of lower labor costs. Similarly, the rise in property incomes relative to wage income could be explained in terms of a perceived need to maintain high real interest rates to limit capital outflows, but it may also owe a good deal to the decision to rely more heavily on incentives to private investors as the main engine of growth.

The longer-term trend of income distribution depends very much on what is done through public programs to lessen remaining inequalities of access to education and skills, to improve public health services and provide better nutrition for the very poor so that they can participate more fully in the production process and take better advantage of new opportunities, and to use the tax system to reduce inequalities rather than to protect them. The net result is not a foregone conclusion: it depends on national choices of economic and social policies. That dependence points to the importance of questions on the second level noted above, the effects of economic liberalization on the range of policy choices actually open to the government.

CAPACITY TO RESPOND TO PUBLIC PREFERENCES

The very nature of an option for liberalization in the broad sense is to reduce the range of choices open to the government. That could be seen as either a gain or a loss for efforts to reduce inequality: intervention has included measures adverse for equality and for democracy, as well as measures going the other way. To give up all kinds of economic and social intervention is probably unthinkable for any government, so what matters is which kinds are eliminated or changed and which are not. Perhaps the fundamental question is the degree to which options to rely on private markets and private investment may bias the pattern of economic policy toward support for property owners and investors, against actions in favor of lower-income groups. A state that relies on incentives for private investors as the main driving force of the economy cannot reproduce the orientation of a state more concerned with a corporatist, inclusionary style of development.

The common presumption that economic liberalization will improve the possibilities for democracy, based on experience in the northern industrialized countries, is misleading for most Latin American countries because of the crucial differences in their economic structures and their implications for inequality. But it certainly has some validity despite these differences. One important positive factor is that liberalization should lessen the monopoly powers and the profits of highly protected firms, through the joint impact of competing imports and entry by new firms. That should lessen the politically based advantages of some of the wealthy and weaken the ties between particular firms and the government officials who make decisions on degrees and kinds of protection. The business and financial community could well remain closely linked to the particular political party that best serves their interests, but this is true everywhere and need not preclude meaningful democracy. The close ties of a single dominant party with a noncompetitive business and financial community that have characterized Mexico since 1940 could be loosened, especially if competing political parties are in fact allowed to win the presidency and prove able to manage economic strategy successfully.

In general, liberalization favors decentralization of power: it undercuts the use of economic controls to fortify state authority and fosters a more complex, pluralistic society, less likely to tolerate authoritarian government. A striking specific example in Mexico of economic controls used to hold down criticism of the government has been the use of a state firm, with a monopoly on production and on imports of newsprint, as an instrument to deny supplies to any publisher considered to be overly independent (see Camp 1989: 205). Freedom of speech and public access to information were constrained by economic controls; removal of this particular restriction by trade liberalization could be an important step toward more genuine democracy. More generally, the omnipresence of regulatory intervention requiring firms to seek discretionary approval for many of their activities, and giving them opportunities to block competition through state action, must have added considerably to the power of the party in control of the state to maintain its unilateral control. On that side of the matter, economic liberalization should favor democracy, if not necessarily a more egalitarian society.

The other side of the coin is that if the state cannot use intervention to control private business decisions, it has less scope to use policies favorable for equity. Eventual improvement of income distribution is less a question of liberalization itself than of such fundamental factors as the relative growth of labor supply and employment opportunities, public investment in human capital, the structure of taxation, the quality of social programs to alleviate poverty, and the rate and stability of economic growth. The question is whether liberalization favors or impedes positive action in such dimensions. Beyond the technical issues discussed above with respect to effects on industrial growth and structure, it involves issues of national goals. A kind of liberalization that gives priority to measures meant to favor foreign investment and to provide strong incentives for private investors can lead to preferences for state action to inhibit labor organization and repress wages, for a low level of taxation that bears more on the general public than on incomes from business and property ownership, for tight restrictions on social programs, including public education, because they cannot be adequately financed under such a tax regime, and against the use of subsidies directed to alleviating poverty because they constitute intervention in the price system and imply the need for greater taxation.

Such an orientation of public policy is not intrinsic to decisions to free international trade, but it has been strongly favored by the pressures of the international financial community on Mexico since the beginning of the debt crisis, especially as expressed by the Bretton Woods institutions and the United States government (Sheahan 1991). Questions of external impacts on domestic policy choices, especially the influence of foreign investment and external creditors, have been debated for a long time as issues of dependency theory. Without trying to resurrect those far-ranging arguments, it is intriguing to see that liberalization and privatization in the style of the Salinas administration constitute a near-total rejection of all the doubts brought out by dependency analysis. If not qualified in practice, the strategy could rule out any independent national policies with respect to relative prices, any lines of investment inconsistent with comparative advantage as determined by world markets interacting with Mexican resources, any independent choice of interest rates, any tax policy considered so adverse by private investors that they react by moving capital out of the country, any social programs considered to threaten property rights or future taxes adverse to business, or any restrictions on advertising and public relations designed from outside to alter Mexican consumer preferences or public opinion. How much meaning would be left to Mexico as an independent society?

A positive answer to this question might be that what would be left would be a more competitive, more efficient society better able to achieve its goals out of more sustained growth of income. Thoroughgoing liberalization would work against some options of public policy but chiefly against the kinds of options adverse to efficiency and growth. Liberalization would mean that excess spending, increases in wages beyond growth of productivity, arbitrarily low interest rates, or currency overvaluation would drive down production and investment so rapidly, and could lead so readily to a massive capital outflow, that populist-style economic policies could not be maintained. Liberalization cuts down the scope for prolonged mistakes.

Such a positive interpretation of the reduced scope for independent national policies fits the logic of neoclassical economic analysis well. That logic is universalist. It gives no value to the strength or distinctive character of a nation. It gives no more value to Mexican than to foreign ownership of productive resources, or to learning by Mexicans than to learning by anyone else able to contribute to the creation of income. It gives no value to democracy, to lessened inequality, to learning by government itself, or to greater variety of preference patterns, except to the degree to which such factors contribute to marketable output.

This antinational strand in open-economy logic runs squarely against long-standing Mexican insistence on strengthening the sense of Mexico as a nation, as a part of the world but a highly distinctive and independent part. The nationalistic side frequently ignores requirements of economic consistency and has cost Mexico dearly; the liberalizing side downgrades concern for national values, and that could have its costs for Mexico as well as for hopes of a diverse world.

If democracy, greater equality, and the development of independent national capacity for entrepreneurship and learning are among the society's goals, the neoclassical strand of economic analysis needs to be complemented by active attention to the broader concepts of development economics and the classical analysis underlying them. Adam Smith was very much concerned with efficiency but also with the wealth of nations. Development economics comes in many versions, some of them unduly minimizing concern for efficiency and macroeconomic balance, but at its best it gives value both to these concerns and to the individual nation's potential to strengthen its own capacities for growth and for independent decision making. That latter side seems to have lost weight in Mexican economic strategy in the process of moving toward liberalization.

CONTRASTS BETWEEN LIBERALIZATION IN MEXICO AND IN OTHER LATIN AMERICAN COUNTRIES

Economic liberalization in the Southern Cone in the 1970s, as in Brazil in the 1960s, was consistently associated with violent political repression. Free markets emphatically did not mean free societies. The association could hardly be considered fortuitous, either in terms of the immediate contexts or in those of longer-run Latin American experience.

The immediate context, in all four of these countries, was one in which economic liberalization was imposed by military governments shortly after the military had overturned democracies following populist or Marxist programs. These programs had all included high degrees of intervention to change the distribution of income, and in Brazil and Chile to alter property rights, at the expense of the higher-income minority. In practice, they also had the effects of driving up inflation, worsening external deficits, and disrupting growth. The military coups overturning these governments were aimed at restoring order and curbing all such intervention in the economy: political repression and redirection of economic strategy came together naturally as a joint rejection of the preceding democratic governments and their economic orientation.

History allowed one counterexample, if the much more modest shift away from protection in Colombia from 1967 can be considered as a somewhat similar move toward liberalization. The immediate context was totally different: a newly elected government chose to reorient national economic strategy in the belief that the changes would be favorable for economic growth as well as for employment. There was no need to call in the military to suppress existing democratic institutions. The reorientation proved to be adverse for real wages for several years, as discussed above, but it also proved to be consistent with gradual reduction of inequality as well as stronger economic growth. In terms of overall economic performance, the results were markedly superior to those registered under the more drastic liberalization programs in the Southern Cone (Foxley 1983; Corbo, de Melo, and Tybout 1986; Ramos 1986).

The exceptional Colombian experience suggests what might have happened if the Brazilian and Southern Cone opposition to their democratic governments had been willing to wait for scheduled elections in order to reorient economic strategies within the existing institutional framework. That course could have saved many lives. It would also have required negotiation and compromise, and surely more qualified liberalization strategies. The Brazilian and Southern Cone reactions were the responses of polarized societies, in which neither property owners nor proponents of radical change sought compromise. The former did not consider democratic institutions to be safe for their futures, and proponents of more egalitarian societies did not consider free markets to be consistent with significant change for the better.

The confrontational character of the four violent cases was not an accident of the immediate context; it was a product of basic characteristics of their economic and political systems. The open market systems generally followed up to the 1930s had long demonstrated both high degrees of inequality and severely restricted scope for national autonomy. When the depression of the 1930s helped break down the political dominance of the groups in favor of continued open systems, opening the way to import-substitution strategies and rejection of reliance on private market forces in general, the Southern Cone countries and Brazil became the regional leaders of extensive state intervention. None of these countries allowed much scope for a proliberalization majority, in part because of the strength of antimarket ideology but basically, underneath that strength, because such a strategy had never been seen to be in the interest of the majority. Opposition to such intervention came to seem necessarily tied to suppression of democratic institutions.

Nineteenth-century Mexico shared many of the economic characteristics giving rise to high inequality under open market conditions. It took the authoritarian rule of Porfirio Díaz to implement and protect a liberal economic strategy. But the aftermath of the Revolution opened up a path different from both the rare democracies in the region and the repressive regimes. It did not open the way to a democracy, to a strongly independent labor movement, or to equality, but instead to an extraordinarily well-organized, inclusionary kind of corporatism. Labor and peasants were effectively incorporated into the dominant political party in the early years of post-Revolution stabilization. Through the 1920s and especially the 1930s, organized labor became both an essential part of the PRI and a major beneficiary in terms of government support for wages, improved working conditions, and social programs (Collier 1981). That method of handling labor relations involved considerable intervention in labor markets, costly special privileges for particular groups (notably for the oil workers' union), and a bias in social investment toward organized urban workers at the expense of the poorer rural sector.[16] It also involved increasing dependence of official organized labor on decisions by the government: when the favors and protection were cut back in the 1980s and real wages were driven down to a degree comparable to the most drastic of the Southern Cone cases, the unions proved to have little capacity, or appetite, for strong reaction.[17] The government was able to carry through liberalization, and an attendant radical change in earnings of labor relative to property, without any significant threat of violent confrontation with the labor force.

The liberalization episodes in the Southern Cone were associated with much more than repression of organized labor: these redirections of economic strategy were accompanied by suppression of all significant forms of democracy and of protection of the individual from violence by the state. The Mexican govenment did not need, and did not choose, to follow that brutal course. The main reason it did not need to do so was that the PRI, through its control of the government, already had control of the election process and was in a position to make its own decisions about how votes are counted. Beginning in the 1970s, electoral reforms began to bring the political system closer to democracy: both left and right opposition could be expressed freely through competing political parties. In the presidential election of 1988 the main party of the left could make clear its opposition to economic liberalization and was able to gain a great many votes in response. But exactly how many votes it gained was never clear because the government found it

---

[16] On the amazing network of special privileges gained by the oil workers' union and the eventual armed confrontation between the government and the head of the union, see Randall 1989.

[17] See especially the analysis of union behavior in the 1980s by Middlebrook (1989).

necessary to interrupt the count until a solution acceptable to the PRI could be negotiated.

There is no need to take away the forms of electoral democracy if, in a political system so dominated by the presidency, opposition parties are not allowed to win the presidential election in any case. It is difficult to believe that the Mexican government would have resorted to direct repression even if the opposition vote had been so overwhelmingly dominant that the result could not have been left uncertain. In that hypothetical case, Mexico would almost surely have entered a new world in which another party could take office, and liberalization would have been sharply curtailed without the kind of bloodshed of the Southern Cone cases. The system does not guarantee peaceful control; it just makes effective challenge extraordinarily difficult. But if the key to such control is to block genuine democracy, that takes the issue back to the basic question: are liberalization and genuine democracy compatible in the Mexican context?

One way that they might be made compatible would be to correct, or at least work toward correcting, the main forces that have made liberalization give such inequitable results. If it actually proved to be beneficial for the majority, the conflict could be resolved. That could be done in part by measures completely consistent with liberalization, most importantly by providing more nearly equal access to education of decent quality, along with better coverage of public health and direct nutritional aid for the poor and a more progressive tax system. It could be done more powerfully by adding on carefully defined departures from free market policies, such as providing price support for peasants producing basic foods, keeping the existing system of minimum wage protection and using it to prevent the minimum from falling in real terms, providing special help for small business and small farmers, applying direct restraint to limit price increases in noncompetitive markets, and fixing exchange rates at levels favorable for sustained growth of industrial exports even if this is not the outcome of existing market forces. All such steps counter to market forces could raise familiar problems if not carefully guided by attention to macroeconomic balance and to minimizing interference with productivity and technological change. If guided carefully, they need not cause any more severe problems than they do in the many industrialized countries which use them with high frequency.

The new social programs of the Salinas government discussed in the preceding section might be seen as an effort to make the process of liberalization less painful to the poor and more acceptable to the majority. They can also be seen as a means of buying political support and criticized for depending on selective programs that do not get at the heart of unequal access to education and resources. They make for a mixed picture in terms of the prospects of democracy. If they allow the PRI to win honest elections, they may favor electoral reforms but at the same time reinforce the problem of a single dominant political party, at least for the time being. Genuine democracy requires greater constraints on the use of state resources to strengthen a particular political party. If such restraint can be established, and an honest electoral system as well, the PRI may still win many elections but the consequences would be different. The difference would be that periods of supremely bad management or highly inegalitarian results would at last be likely to bring an opposition party into power. Economic liberalization might be qualified greatly in the interests of maintaining political support, or it might not need to be so qualified if in fact it yields positive results for the majority.

ECONOMIC LIBERALIZATION IN THE ABSENCE OF POLITICAL LIBERALIZATION

While it is distinctly possible that Mexico could adopt full political liberalization in the near future and stick to it, pressures remain strong for the PRI to retain safety-first restrictions to ensure continued control of the presidency. The pressures come from a long-established institution that has served many people inside the system well, from both business and labor leadership accustomed to negotiation with a stable governing party, and more generally from fear of the unknown. The growing strength of countering forces in favor of more meaningful democracy is evident as well, but they could conceivably be held off for a long time by such a well-entrenched, and in most periods adept, political organization. What would it mean for the course of economic liberalization if political liberalization is indefinitely postponed?

In the near term, the most likely effect is that political control could be used to protect economic liberalization from pressures to amend it to meet popular preferences. If the PRI had lost the presidency in 1988 there seems to be little doubt that economic liberalization would have been greatly qualified, if not rejected. For those who favor economic liberalization, political liberalization is a danger.

The appeal of economic liberalization as an alternative to the biased and sometimes costly kinds of intervention practiced under protection in the past is real enough and has clearly won wide support. If it is given a sustained trial and actually proves to be consistent with improving standards of living for the great majority of the country, then it could probably go on indefinitely without threat from political liberalization. For many people, this possibility argues in favor of taking one step at a time. It probably underlies the frequent suggestion that it is too much to expect any society to carry out a high degree of political liberalization at the same time as economic liberalization: that it is more realistic, and safer, to concentrate on the economic side first.

This conception of necessary sequence can be supported by history but may be costly for Mexico's future. Nineteenth-century economic liberalism in England and the United States developed under limited democracies that initially restricted voting rights for lower-income groups, in a process that both favored economic growth and allowed gradual movement to more complete democracy (Persson and Tabellini 1991). In postwar Chile and Spain, economic liberalization was imposed by authoritarian governments but both were able to move to democracy later without, at least so far, repudiating liberalization. And the strains being experienced in Eastern Europe as so many countries try to achieve economic and political liberalization simultaneously point to the same interpretation. But it is not clear what such examples mean for Mexico. Should they be taken to mean that pursuit of economic liberalization both requires and justifies postponement of democratization? Or perhaps instead that the particular kinds of liberalization should be guided by concern for the welfare of the great majority of the country, in order to be consistent with democracy? The danger of opting to postpone democracy for the sake of liberalization is evident: choices of economic strategy geared primarily to encourage private investment, foreign and domestic, can work against the kinds of structural change needed for lessened inequality and thereby against, rather than for, the longer-term chances of democracy.

The dangers of accentuating inequality through liberalization could be countered by a recognition of multiple goals, of the need for active state leadership to widen access to education and skills and to use the fiscal system to limit inequality. Such countering action would be likely in an open and informed democracy, in which public preferences could place limits on the particular goal of favoring the business sector to encourage private investment. That is perhaps the heart of the question. If Mexico were an open and informed democracy, all-out liberalization in the sense of withdrawal from state guidance of the economy in favor of reliance on private markets would be subject to the kinds of social constraints and direction more favorable for equality. Liberalization would be limited and to some extent offset in order to further social goals that go beyond economic efficiency and growth.

Some Implications and Questions for Latin America

If Mexico both perseveres in its strategy of liberalization and manages to demonstrate important gains as a result, the tentative moves in that direction in other countries of the region will surely gather strength. Mexico, along with Chile and Colombia but with perhaps more impact, could lead the region toward more extensive economic liberalization in much the same way that Argentina and Brazil led it toward import substitution fifty years ago. That should help clear up the particular problems caused by import substitution though it may not do much to correct, or may even aggravate, the underlying factors that make for such high degrees of inequality. Perhaps most importantly, the changes could go far deeper than reorientation toward more open economies, with less government regulation and perhaps greater economic efficiency. They would foster values and responses that run contrary to much of Latin American history.

Exactly what kinds of changes are likely to spread as a consequence of Mexico's new economic strategy depend mainly on each country's own constellation of political forces but also on what forms the Mexican strategy takes. Mexico does not need to copy either Pinochet's Chile or the reserved kind of liberalization of Colombia. It can go somewhere in between these versions or create a distinctive model of its own. The options are highly important for their social as well as their economic consequences in Mexico, as discussed above. They are also important for the rest of Latin America, insofar as the consequences in Mexico influence choices in the other countries.

Three likely influences of the Mexican path are considered briefly here: (1) new balances in the kinds of economic policies likely to be considered and adopted, particularly going against heterodox or populist macroeconomic strategies, against regional agreements to establish common strategies for such purposes as debt negotiation, control of foreign investment, or controls on capital movements; (2) changing possibilities for democracy, in either positive or negative directions, depending in part on the economic structure and particular version of the strategy in each country; and (3) changing attitudes and social identification, away from corporatist solutions and toward competitive individualism, and away from nationalism and regionalism toward identification with "the north."

The kind of liberalization initiated in Mexico in the second half of the 1980s should be expected to cut down considerably on the likelihood and the duration of experiments with populist economic polices. An open economy does not leave much room, or time, for policies that give rise to strong domestic inflation and external deficits. It would of course remain possible for a government determined to implement populist policies to revert to protection and overvaluation: with liberalization reversed, a populist program might go for several years without collapse. But if the basic strategy of liberalization remains in place, it could not last. The costs would come swiftly and much more visibly to all the interest groups concerned. And if the particular country's problems stood out against greater stability in the rest of the region, the political forces against populism would be strengthened and those in favor weakened. An instructive example from the more liberalized Europe of the EEC is the fate of President Mitterand's short-lived experiment with socialist measures in France at the beginning of the 1980s: the economic costs showed up so swiftly that the Socialist party almost immediately reversed its economic strategy to follow much more conservative policies as the perceived route to political survival.

Similarly, liberalization works against regional agreements to establish collective defenses of any separate course of action. If it extends to investment and other capital movements, it would practically rule out such possibilities as a debtors' cartel, or independent efforts to enforce interest rates below those prevailing in the outside world, or taxation of returns to capital that brought them significantly below returns in external financial markets. Liberalization to any high degree means integration with the outside, denying barriers and undercutting differences.

It might be considered that turning away from regional agreements or limits on capital movements would not represent any great change from the past, given the repeated weakness of Latin American attempts to establish joint negotiating positions on such matters as external debts or to impose effective limits on capital movements. But these are questions of degree, and whatever the possibilities in the past, they would be reduced by liberalization.

This may be especially important with respect to capital movements: a few countries have been able to control them more effectively than others and thus achieve somewhat greater degrees of economic policy independence (Williamson and Lessard 1987). Serious liberalization would remove this support for autonomy. It would change the balance of possible choices in favor of greater independence of investors from the government, and greater need for governments to weigh the preferences of private investors in making decisions on all issues of economic policy.

The implication of liberalization that governments that stick to it must give greater weight to the preferences of private investors has important bearing on the second set of questions: how any thoroughgoing version of liberalization may affect the possibilities for sustained democracy. But it is easier to see that the connections are significant than to be confident about how the balance would work out. One group of connections should be favorable for democracy and a second group unfavorable.

Economic liberalization could support an existing democracy by favoring more successful economic growth, raising collective confidence in the capacity of the society for overall performance, and providing more national income to respond to multiple claims. Further, if it tilts the balance of economic and social policies toward greater concern for private investors, the repressive right-wing military coups so frequent in the past may be seen by the conservative side of the society as less necessary, at the same time as they are seen by the rest of the society to be less likely. If Latin American democracies were typically overturned from the left, the changed balance could make overturns more probable; since they have more often been cut short by coalitions of middle- and upper-income groups with the military, the changed balance should make this common form of overturn less probable. The cost, unhappily, is that the societies could become even more inequitable than in the past.

It is the implication of economic liberalization for poverty and inequality that makes any all-out version problematic for sustained democracy in Latin America. With initially concentrated wealth, excessively abundant and rapidly growing unskilled labor outrunning productive employment opportunities, and national economic policy constrained even more than in the past by the need to avoid measures perceived to be adverse for private investors, the outlook for income distribution is not promising. The outlook could be improved by better macroeconomic management to foster a more stable and more employment-creating kind of economic growth, by more active social programs in education, nutrition, and public health, and by more effective and more progressive tax systems. These are real possibilities. The problem with economic liberalization is that it could shift the balance of policy concern and the political weight within these societies even more than in the past toward the objectives of private investors, running counter to the likely social goals of a democratic society.

Given these two-way implications for the effects on democracy, the potential gains from liberalization could be more dependably secured if the economic structures responsible for such high inequality are attacked by specific corrective policies within a gradual process of liberalization. The more progress that could be made by corrective economic policies, by active state efforts to change the distribution of gains in the direction of lessened inequality, the more thoroughly liberalization could be carried through. The process of liberalization and that of structural change could complement each other, toward goals that go beyond improved economic efficiency to lessened inequality and sustainable democracy.

Sustainable democracy is a question of cultural values, attitudes, social relationships, and the balance of power among conflicting interest groups; not in any direct sense a question of economic strategy. Still, changes in economic strategy can shift the odds by lessening or worsening the kinds of economic strains that can tear societies apart, and perhaps beyond that by affecting social values and relationships. The third set of questions concerns the possible ways in which the new economic strategy may encourage changes in social attitudes and in group perceptions of their own interests.

The elusive connections between economic strategies and a nation's social and political relationships make it easy to go wrong. Albert Hirschman suggests that insistence on necessary conditions for democracy, such as dynamic economic growth or improved income distribution, can undermine its survival by discrediting it. Specific demands considered essential can discourage any willingness to negotiate to save a democratic system when it is under strain. "Preconditions" for democracy are futile: "they would merely serve to spell out a wholly utopian scheme for changing everything that has been characteristic of Latin American reality, and would therefore amount to wishing away that reality" (Hirschman 1986: 176).

Hirschman's insight applies both to economic liberalization and to the pursuit of democracy in Latin America: the issues involve "everything that has been characteristic of Latin American reality." Within the intricate and often conflicting strands of that reality, it would change the balance between concern for the importance of the society as a whole relative to individual profit seeking, against traditional relationships of mutual obligation between individuals and their extended families and communities, and against conceptions of the responsibility of the state to promote an inclusive social order. These more traditional values all have their positive and their troublesome sides. They can lend support to authoritarian regimes conceived as the best way to maintain social order, and they can also lend support to socially concerned state action to put help for the weak ahead of economic efficiency. They can support restriction of individual freedom, and they can open up life opportunities for those who would have been condemned to poverty in a more individualistic society.

The likelihood that pressures in favor of economic liberalization would foster changes in the fundamentals of Latin American reality could be seen either as a promise of a better future or as a threat of serious loss. Or perhaps both at once, in a mixture difficult to predict. Latin American reality has not in general been favorable for its poor, and that means a high share of its people. It has not in general been favorable for equity, internal political peace, sustained economic growth, or social integration. It has been favorable for many other human and cultural values, so fundamental change could bring painful loss as well as gain. A great deal is at stake, in many different directions.

Given the possibility that thoroughgoing economic liberalization could foster significant changes in Latin American reality, and that the changes could be both positive and negative, arguments over its desirability should surely not be restricted to its economic effects. For the same reason, pressures from outside to drive Latin America toward greater liberalization—particularly from the international financial agencies and the government of the United States—would seem to verge on a peculiarly sensitive kind of interventionism. To argue in favor of liberalization is one thing, but to deny or grant needed economic support on condition of going further in this direction is something considerably more serious, and more dubious.

The crucial question in this context is, what is the balance of goals among Latin Americans themselves? Do most of them want to change the orientation of their societies in the ways that economic liberalization would favor? Do they want to change the balance away from concern for integrating their own societies, toward greater emphasis on integration of the more modern side of each country with the outside world? Do they want to become less distinctive and more worldly? The region's trend toward democracy in the last decade may permit broader participation than ever before in reaching decisions on the specific policies that will shape the course of change. Can the outside world—the international financial institutions and creditor governments in particular—acquire the patience to allow the Latin Americans themselves to work out their own decisions on the degrees and kind of liberalization that they find consistent with their own majority preferences?

Conclusions

Multiplicity of goals and many uncertainties complicate interpretation of Mexico's response to the debt crisis and its dramatic redirection of economic strategy. The new strategy has set powerful forces in motion, some clearly positive and some probably not. Perhaps the clearest conclusions are that the country has found a costly but so far relatively successful way to move past the debt crisis, that the structure of the economy is changing in fundamental ways that should be favorable for efficiency but probably not for equity, and that economic liberalization raises both hopes and problems for the prospects of more meaningful democracy.

Mexico's conservative response to the debt crisis cost the country a great deal in terms of output per capita, investment, and living standards for the lower-income majority. It probably aggravated an initially high degree of inequality. The results for output and wages look weak compared to those of Chile and Colombia in the same period, although much better compared to the more unorthodox approaches of Argentina, Brazil, and Peru. With respect to inflation, Mexico seemed to be losing control badly at first, much like the latter three countries, but then managed to break away from their unsuccessful path and match the degree of restraint achieved in Chile and Colombia.

In terms of success with developing more competitive industrial exports, Mexico did far better than any of these other five countries. The main keys to that success were preceding administrative measures to force selected import-substitution firms to invest in production for new exports, high degrees of devaluation of the real exchange rate to make exports more profitable, tight restraint of the primary fiscal deficit, liberalization of trade to reduce costs of imported inputs and to increase domestic competition, and from 1987 a distinctive form of incomes policy. In a most impressive balancing act, promotion of new exports has been combined with improvement of restraint on inflation, two goals that are almost inescapably in conflict. Perhaps the most troublesome immediate question is the course of real exchange rates and thereby of the ability to maintain industrial growth in an open economy: from 1988 through 1990 the real exchange rate was allowed to appreciate considerably in the interest of stopping inflation, to a point at which continued ability of the industrial sector to compete could be undermined.

Liberalization helped to restrain inflation and to encourage exports, but more importantly it has become a leading goal in its own right. It has come to include the practical elimination of restraints on foreign investment and other forms of capital movement, as well as a generalized withdrawal from the whole system of government control and support that has maintained Mexico's relatively inclusionary corporative system ever since the 1930s. It could encourage renewed growth by giving private investors more freedom and stronger incentives. But by the same token it undercuts the power of the state to take a positive role in shaping the economy. Collective social preferences count for less relative to those of individual investors, and the possibilities of government leadership toward a more equitable society are reduced.

These fundamental changes are bound to give Mexico a different kind of society, in some respects probably more dynamic but in other respects possibly even more inequitable than in the past. The main direct concern for equity is that the changes are likely to favor incomes of property owners and the highly skilled relative to incomes of unskilled urban and rural labor. The more general concern is that they raise the economic and political weight of private investors relative to other group interests in the society, and thereby reduce the chances of public action to change the basic structural conditions of the Mexican economy that have been so unfavorable for equality in the past.

Under these conditions, liberalization and popular preferences run contrary to each other. This conflict raises the danger that Mexican governments may continue to limit democracy in the interest of protecting the new economic strategy. Thorough democracy is a big risk. To some of those concerned mainly with better economic performance, it seems excessively risky to pursue political and economic liberalization simultaneously. The danger of trying to do too much at once is evident. But then so is the danger that holding off democracy to protect the new economic strategy will favor a more inequitable kind of liberalization than necessary and will build up resistance to the kinds of changes needed to correct the structural conditions adverse to equality.

Conclusions

If Mexico pursues economic liberalization in ways able to gain and hold popular acceptance, the same redirection of economic strategy is likely to become widespread in Latin America. That would further reduce whatever possibilities now exist to control foreign investment, to maintain differential relative prices in the interest of lessening poverty and inequality, or to insulate the region from external economic influence in any other ways. It could also mean increased efficiency and output, lessened dangers of runaway inflation, and greater ability to avoid foreign exchange crises. But all of these consequences depend on exactly how liberalization is implemented, in what context of accompanying economic and social policies, subject to what kinds of external constraints and pressures. The strong pressures being exerted by the international financial agencies and the United States in favor of early and thorough liberalization carry dangers for Latin American autonomy and for democracy. They risk aggravating inequality, heightening social conflict, and undermining the confidence of Latin Americans to negotiate viable economic strategies acceptable to their own majorities.

Under economic conditions common in Latin America, thoroughgoing forms of economic liberalization can have adverse effects for the low-income majority, and would therefore not be the likely outcome in a fully democratic society. For those confident that economic liberalization is highly desirable, that conflict can be seen as a reason for delaying or restricting political liberalization. But it can also be seen as a reason for the opposite course, for giving priority to political liberalization so that the majority of the country can participate in determining the particular forms of economic liberalization they prefer. If democracy is truly one of the goals, it should be brought to bear on the design and implementation of any liberalization strategy, rather than held off while the economy is restructured toward more exclusive reliance on market forces.

References

Ahluwalia, Montek S. 1974. "Income Inequality: Some Dimensions of the Problem." In *Redistribution with Growth: An Approach to Policy*, edited by Hollis B. Chenery et al. Oxford: Oxford University Press, for the World Bank.

Ahluwalia, Montek S., Nicholas G. Carter, and Hollis B. Chenery. 1979. "Growth and Poverty in Developing Countries," *Journal of Development Economics* 6.

Alberro, José, and Jorge E. Cambiaso. 1989. "Características del ajuste de la economía mexicana." In *Políticas económicas y brecha externa: América Latina en los años ochenta*, edited by Nicolás Eyzaguirre and Mario Valdivia. Santiago: Comisión Económica para América Latina y el Caribe (ECLAC).

Austen, James E., and Gustavo Esteva, eds. 1987. *Food Policy in Mexico: The Search for Self-Sufficiency*. Ithaca, N.Y.: Cornell University Press.

Banco de México. 1990. "Economic Indicators." Mexico: Banco de México, September.

Bazdresch, Carlos, and Santiago Levy. 1990. "Populism and Economic Policy in Mexico, 1970–1982." Discussion Paper No. 12. Boston: Boston University Institute for Economic Development, May.

Behar, Jaime. 1988. *Trade and Employment in Mexico*. Stockholm: Swedish Institute for Social Research.

Beristain, Javier, and Ignacio Trigueros. 1990. "Mexico." In *Latin American Adjustment: How Much Has Happened?* edited by John Williamson. Washington, D.C.: Institute of International Economics.

Bourguinon, François, and Christian Morrison. 1989. *External Trade and Income Distribution*. Paris: Organisation for Economic Co-operation and Development.

Camp, Roderic. 1989. *Entrepreneurs and Politics in Twentieth-Century Mexico*. Oxford: Oxford University Press.

Cárdenas, Enrique. 1987. *La industrialización mexicana durante la gran depresión*. Mexico: El Colegio de México.
———. 1990. "Contemporary Economic Problems in Historical Perspective." In *Mexico's Search for a New Development Strategy*, edited by Dwight S. Brothers and Adele E. Wick. Boulder, Colo.: Westview.
Collier, Ruth Berins. 1981. "Popular Sector Incorporation and Political Supremacy: Regime Evolution in Brazil and Mexico." In *Brazil and Mexico: Patterns in Late Development*, edited by Sylvia Ann Hewlett and Richard Weinert. Philadelphia: Institute for the Study of Human Issues.
Corbo, Vittorio, Jaime de Melo, and James Tybout. 1986. "What Went Wrong with the Recent Reforms in the Southern Cone?" *Economic Development and Cultural Change* 34 (April).
Cortázar, René. 1980. "Distribución del ingreso, empleo y remuneraciones reales en Chile, 1970–78," *Colección Estudios CIEPLAN* 3 (June).
Cottani, Joaquín A., Domingo F. Cavallo, and M. Shahabaz Khan. 1990. "Real Exchange Rate Behavior and Economic Performance in LDCs," *Economic Development and Cultural Change* 19:1 (October): 61–76.
Díaz Alejandro, Carlos. 1983. "Open Economy, Closed Polity?" In *Latin America in the World Economy*, edited by Diana Tussie. New York: St. Martin's.
Dresser, Denise. 1991. "Neopopulist Solutions to Neoliberal Problems: Mexico's National Solidarity Program." Current Issue Brief Series, no. 3. La Jolla: Center for U.S.-Mexican Studies, University of California, San Diego. Forthcoming November.
ECLAC (Economic Commission for Latin American and the Caribbean). 1990. *Preliminary Overview of the Economy of Latin America and the Caribbean, 1990*. Santiago: ECLAC, December.
Edwards, Sebastian. 1989. *Real Exchange Rates, Devaluation, and Adjustment: Exchange Rate Policy in Developing Countries*. Cambridge, Mass.: MIT Press.
Feres, Juan Carlos, and Arturo León. 1990. "Magnitud de la situación de la pobreza," *Revista de la CEPAL* 41 (August).
Foxley, Alejandro. 1983. *Latin American Experiments in Neoconservative Economics*. Berkeley: University of California Press.
Grindle, Merilee S. 1991. "The Response to Austerity: Political and Economic Strategies of Mexico's Rural Poor." In *Social Responses to Mexico's Economic Crisis of the 1980s*, edited by Mercedes González de la Rocha and Agustín Escobar Latapí. Contemporary Perspectives Series, no. 1. La Jolla: Center for U.S.-Mexican Studies, University of California, San Diego.

Helpman, Elhanan, and Paul Krugman. 1985. *Market Structure and Foreign Trade*. Cambridge, Mass.: MIT Press.

Hirschman, Albert. 1979. "The Turn toward Authoritarianism in Latin America and the Search for Its Economic Determinants." In *The New Authoritarianism in Latin America*, edited by David Collier. Princeton, N.J.: Princeton University Press.

———. 1986. "Notes on Consolidating Democracy in Latin America." In *Rival Views of Market Society*, edited by A. Hirschman. New York: Viking.

IDB (Inter-American Development Bank). 1990. *Economic and Social Progress in Latin America, 1990 Report*. Baltimore, Md.: Johns Hopkins University Press for the IDB.

———. 1991. *Economic and Social Progress in Latin America, 1991 Report*. Baltimore, Md.: Johns Hopkins University Press for the IDB.

IMF (International Monetary Fund). 1983. *International Financial Statistics, 1983 Yearbook*. Washington, D.C.: IMF.

King, Robin A. 1990. "Determinants of Macroeconomic Policy under External Debt Crisis: Mexican Debt and Adjustment Policy, 1981–1990." Austin: Department of Economics, University of Texas at Austin, November.

Krugman, Paul. 1980. "Scale Economies, Product Differentiation, and the Patterns of Trade," *American Economic Review* 70:5 (December): 950–59.

Levy, Santiago. 1991. "Poverty Alleviation in Mexico." Working Paper No. 679. Washington, D.C.: Policy, Research and External Affairs, World Bank.

Lustig, Nora. 1989. "The Impact of the Crisis on Living Standards in Mexico: 1982–1985." Helsinki: World Institute for Development Economics Research, June.

Mercado, Alfonso, and Taeko Taniura. 1990. "The Mexican Automotive Export Growth: Favorable Factors, Obstacles and Policy Requirements." Documento de Trabajo No. 9. Mexico: Centro de Estudios Económicos, El Colegio de México.

Middlebrook, Kevin. 1989. "The Sound of Silence: Organized Labour's Response to Economic Crisis in Mexico," *Journal of Latin American Studies* 21:1 (May).

Persson, Torsten, and Guido Tabellini. 1991. "Is Inequality Harmful for Growth?" Working Paper No. 3599. Cambridge, Mass.: National Bureau of Economic Research, January.

Ramos, Joseph. 1986. *Neoconservative Economics in the Southern Cone of Latin America, 1973–1983*. Baltimore: Johns Hopkins University Press.

Randall, Laura. 1989. *The Political Economy of Mexican Oil*. New York: Praeger.

Ros, Jaime. 1991. "Industrial Organization and Comparative Advantage in Mexico's Manufacturing Trade." Working Paper No. 155. Notre Dame, Ind.: Helen Kellogg Institute for International Studies, University of Notre Dame, March.

Ruiz Durán, Clemente. 1991. "Changes in the Industrial Structure and the Role of Small and Medium Industries in Developing Countries. The Case of Mexico." Summary report prepared for the conference on Small and Medium Industries in Developing Countries. Tokyo: Institute of Developing Economies, February.

Sachs, Jeffrey. 1987. "Trade and Exchange Rate Policies in Growth-Oriented Adjustment Programs." In *Growth-Oriented Adjustment Programs*, edited by Vittorio Corbo, Morris Goldstein, and Mohsin Khan. Washington, D.C.: International Monetary Fund and World Bank.

———. 1989. "Social Conflict and Populist Policies in Latin America." Working Paper No. 2897. Cambridge, Mass.: National Bureau of Economic Research, March.

Sheahan, John. 1987. *Patterns of Development in Latin America: Poverty, Repression, and Economic Strategy*. Princeton: Princeton University Press.

———. 1991. "Economic Forces and U.S. Policies." In *Exporting Democracy: The United States and Latin America*, edited by Abraham Lowenthal. Baltimore: Johns Hopkins University Press.

———. n.d. "The Economy." In country study of Peru being prepared by the Library of Congress, forthcoming.

Solís, Leopoldo. 1990. "Social Impact of the Economic Crisis." In *Mexico's Search for a New Development Strategy*, edited by Dwight S. Brothers and Adele E. Wick. Boulder, Colo.: Westview.

Twomey, Michael. 1990. "Neo-Liberal Modernization: The Mexican Automobile Industry." Dearborn: University of Michigan, August.

Unger, Kurt. 1990. *Las exportaciones mexicanas ante la reestructuración industrial internacional: La evidencia de las industrias química y automotriz*. Mexico: El Colegio de México/Fondo de Cultura Económica.

Urrutia, Miguel. 1981. "Experience with the Crawling Peg in Colombia." In *Exchange Rate Rules*, edited by John Williamson. New York: St. Martin's.

———. 1985. *Winners and Losers in Colombia's Economic Growth*. Oxford: Oxford University Press.

Villarreal, René. 1988. *Industrialización, deuda y desequilibrio externo en México: Un enfoque neoestructuralista (1929–1988)*. 2d. ed. Mexico: Fondo de Cultura Económica.

———. 1990. "Del estado de bienestar al estado solidario," *Examen* 2:15 (June 15).

Whitehead, Laurence. 1989. "Political Change and Economic Stabilization: The 'Economic Solidarity Pact.'" In *Mexico's Alternative Political Futures*, edited by Wayne A. Cornelius, Judith Gentleman, and Peter H. Smith. Monograph Series, no. 30. La Jolla: Center for U.S.-Mexican Studies, University of California, San Diego.

Williamson, John, and Donald R. Lessard. 1987. *Capital Flight: The Problem and Policy Responses*. Policy Analysis No. 23. Washington, D.C.: Institute of International Economics, November.

World Bank. 1990. *World Development Report, 1990*. New York: Oxford University Press, for the World Bank.

———. 1991a. *World Development Report, 1991*. New York: Oxford University Press, for the World Bank.

———. 1991b. *World Tables, 1991*. Baltimore: Johns Hopkins University Press, for the World Bank.

Zamosc, Leon. 1986. *The Agrarian Question and the Peasant Movement in Colombia: Struggles of the National Peasant Association, 1967–1981*. Cambridge: Cambridge University Press.

About the Author

John Sheahan is a professor of economics at Williams College. He has worked as an economic adviser in several countries in Asia and in Latin America, spent a year as a researcher and teacher at El Colegio de México, and was a visiting fellow at the Center for U.S.-Mexican Studies at the University of California, San Diego, in the spring of 1991. His special interests are the international economic problems of developing countries, particularly in Latin America, and the relationships between economic strategies and political strains. His latest book is *Patterns of Development in Latin America: Poverty, Repression, and Economic Strategy* (Princeton University Press, 1987).